THE RISE OF OBESITY IN EUROPE

The Rise of Obesity in Europe

A Twentieth Century Food History

Edited by

DEREK J. ODDY
University of Westminster, UK

PETER J. ATKINS
Durham University, UK

VIRGINIE AMILIEN
National Institute for Consumer Research, Oslo, Norway

ASHGATE

Published by
Ashgate Publishing Limited
Wey Court East
Union Road
Farnham
Surrey, GU9 7PT
England

Ashgate Publishing Company
Suite 420
101 Cherry Street
Burlington
VT 05401-4405
USA

www.ashgate.com

British Library Cataloguing in Publication Data
International Commission for Research into European Food
 History. Symposium (10th : 2007 : Oslo, Norway)
  The rise of obesity in Europe : a twentieth century food
  history.
  1. Food consumption--Europe--History--20th century--
  Congresses. 2. Food habits--Europe--History--20th century--
  Congresses. 3. Obesity--Social aspects--Europe--Congresses.
  4. Obesity--Economic aspects--Europe--Congresses.
  I. Title II. Oddy, Derek J. III. Atkins, P. J. (Peter J.)
  IV. Amilien, Virginie.
  306.4'613'094'09045-dc22

Library of Congress Cataloging-in-Publication Data
Oddy, Derek J.
  The rise of obesity in Europe : a twentieth century food history / by Derek J. Oddy, Peter J. Atkins and Virginie Amilien.
       p. cm.
  Includes index.
  ISBN 978-0-7546-7696-6 (hardback) -- ISBN 978-0-7546-9395-6 (ebook)
 1. Food habits--Europe--History--20th century. 2. Food preferences--Europe--History--20th century. 3. Food consumption--Europe--History--20th century. 4. Food supply--Europe--History--20th century. 5. Obesity--Europe--History--20th century. I. Atkins, P. J. (Peter J.) II. Amilien, Virginie. III. Title.

  GT2853.E8O33 2009
  394.1'20940904--dc22

                                                                        2009021342

ISBN 9780754676966 (hbk)
ISBN 9780754693956 (ebk)

Printed and bound in Great Britain by
TJ International Ltd, Padstow, Cornwall

Contents

PART 3: SOCIAL AND MEDICAL INFLUENCES

List of Figures

List of Tables

List of Contributors

Virginie Amilien, National Institute for Consumer Research (SIFO), PO Box 4682 Nydalen, 0405 Oslo, Norway. Email: virginie.amilien@sifo.no

Peter J. Atkins, Department of Geography, Durham University, Durham DH1 3LE, UK. Email : p.j.atkins@durham.ac.uk

Julia Csergo, Université Lyon 2, Laboratoire d'études rurales (EA 3728-Usc Inra 2024); 21 rue du Temple, 75004 Paris, France. Email: julia.csergo@wanadoo.fr

Runar Døving, Professor of Social Anthropology, Oslo School of Management, and National Institute for Consumer Research (SIFO), PO Box 4682 Nydalen, 0405 Oslo, Norway. Email: runar.doving@sifo.no

Alain Drouard, Centre Roland Mousnier, UMR 8596 du Cnrs, Université de Paris – Sorbonne (Paris IV). Mail : 16, rue Parrot, 75012 Paris, France. Email: adrouard01@noos.fr

Andreas Exenberger, Research Area 'Economic and Social History', Department of Economic Theory, Policy, and History, Faculty of Economics and Statistics, Innsbruck University, Universitaetsstrasse 15, A-6020 Innsbruck, Austria. Email: andreas.exenberger@uibk.ac.at

Martin Franc, Masaryk Institute, Archives of the Academy of Sciences of Czech Republic. Gabcikova 2362/10, 182 00 Praha 8, Czech Republic. Email: francmartin@seznam.cz

Maja Godina Golija, Institute of Slovenian Ethnology, SRC SASA, Novi trg 2, 1000 Ljubljana, Slovenia. Email: maja.godina-golija@uni-mb.si

Unni Kjærnes, National Institute for Consumer Research (SIFO), PO Box 4682 Nydalen, 0405 Oslo, Norway. Email: unni.kjarnes@sifo.no

Josef Nussbaumer, Research Area 'Economic and Social History', Department of Economic Theory, Policy, and History, Faculty of Economics and Statistics, Innsbruck University, Universitaetsstrasse 15, A-6020 Innsbruck, Austria. Email: josef.nussbaumer@uibk.ac.at

Derek J. Oddy, University of Westminster, 309 Regent Street, London W1R 8AL, UK. Email: derekjoddy@aol.com

Gun Roos, National Institute for Consumer Research (SIFO), PO Box 4682 Nydalen, 0405 Oslo, Norway. Email: gun.roos@sifo.no

Gloria Sanz Lafuente, Departamento de Economía, Universidad Pública de Navarra, Campus de Arrosadía 31006 Pamplona, Spain. Email: gloria.sanz@unavarra.es

Jürgen Schmidt, Social Science Research Centre Berlin, Historical Social Sciences, Reichpietschufer 50, D-10785 Berlin, Germany. Email: jschmidt@wzb.eu

Hans Jürgen Teuteberg, Westfälische Wilhelms-Universität Münster, Historisches Seminar, Domplatz 20-22, 48143 Münster, Germany. Email: teuteberg-uni-muenster@t-online.de

Ulrike Thoms, Institut für Geschichte der Medizin, Zentrum für Human- und Gesundheitswissenschaften, Humboldt Universität / Freie Universität Berlin, Klingsorstr.119,12203 Berlin, Germany. Email: ulrike.thoms@charite.de

Tatiana Voronina, Institute of Ethnology, Russian Academy of Sciences, Leninsky Prospect 32 A, Moscow, 11 73 34, Russia. Email: russkie2@iea.ras.ru

Ina Zweiniger-Bargielowska, Department of History, University of Illinois, Chicago, USA. Email: inazb@uic.edu

Preface and Acknowledgments

The publication of this volume is a celebration of the twentieth anniversary of the foundation of the International Commission for Research into European Food History. The aim of ICREFH is the study of the history of food and nutrition over the last 250 years and every two years a symposium is organized on a particular theme within that broad remit. Our discussions are normally spread over three or four days, with each session devoted to deliberation around a number of pre-circulated papers. The ICREFH difference is this focus upon debate rather than upon formal presentations, with the result that there is time for comparative insights from the many European countries represented and also cross-cutting perspectives from the different disciplinary standpoints. This intense and productive experience is then published as a selection of peer-reviewed papers.

The hosts for the tenth ICREFH symposium were the Statens institutt for forbruksforskning (SIFO), the National Institute for Consumer Research in Norway. The symposium was entitled 'From Under-Nutrition to Obesity: Changes in Food Consumption in Twentieth-Century Europe' and was held in Oslo from 25th to 29th September 2007. Food consumption and health policy have been major research themes at SIFO for the last twenty years and ICREFH's historical dimension provided an enriching perspective. SIFO is a non-biased governmental institute concerned with consumer research and depending on the Ministry of Children and Family Affairs. Involving nine European countries (Austria, the former Czechoslovakia, France, Germany, Norway, Russia, Slovenia, Spain, and the United Kingdom) using qualitative and quantitative data, the present book throws light on social, economic, industrial, commercial, political, and medical aspects that directly contribute to SIFO's research interests, namely how 'the image of the consumer' materializes and how public and private regulations contribute to the shape and conditions of strategies and practice along the chain of values.

Food and health are central themes for the 'Market and Politics' research group at SIFO that organized the symposium. They particularly focus on the situation of the consumer, by examining different systems, strategies and understandings in the structuring of consumer choices. The argument is that it is necessary to understand what is going on 'behind the counter' in order to understand consumer choices of what is 'on the counter'.

The theme of the symposium was complementary to SIFO's studies on food and health. Already in the 1980s it had published on Norwegian food policy, showing that the country's nutrition policy had been founded on strong historical traditions of state responsibility for public welfare as well as public support for agriculture. This policy has a unique character, even compared to other Nordic countries. SIFO has already taken the idea of comparative research further in a

European Union project, 'Partnership, Healthy Eating and Innovative Governance as Tools to Counteract Obesity and Overweight − Obesity Governance', which began in May 2009. This is a study of six countries, coordinated by SIFO, within the Health Programme's theme of 'Nutrition, Overweight and Obesity-Related Health Issues'.

ICREFH Symposia

The Current State of European Food History Research (Münster, 1989)
The Origins and Development of Food Policies in Europe (London, 1991)
Food Technology, Science and Marketing (Wageningen, 1993)
Food and Material Culture (Vevey, 1995)
Order and Disorder: the Health Implication of Eating and Drinking (Aberdeen, 1997)
The Landscape of Food: Town, Countryside and Food Relationships (Tampere, 1999)
Eating and Drinking Out in Europe since the late Eighteenth Century (Alden Biesen, 2001)
The Diffusion of Food Culture: Cookery and Food Education during the Last 200 Years (Prague, 2003)
Food and the City in Europe since the Late Eighteenth Century: Urban Life, Innovation and Regulation (Berlin, 2005)
From Under-Nutrition to Obesity: Changes in Food Consumption in Twentieth-Century Europe (Oslo, 2007)
Food and War in Europe in the Nineteenth and Twentieth Centuries (Paris, 2009)

ICREFH Publications

ICREFH I – Teuteberg, H.J. (ed.), *European Food History: a Research Overview*, Leicester, 1992.
ICREFH II – Burnett, J. and Oddy, D.J. (eds), *The Origins and Development of Food Policies in Europe*, London, 1994.
ICREFH III – Hartog, A.P. den (ed.), *Food Technology, Science and Marketing: European Diet in the Twentieth Century*, East Linton, 1995.
ICREFH IV – Schärer, M.R. and Fenton, A. (eds), *Food and Material Culture*, East Linton, 1998.
ICREFH V – Fenton, A. (ed.), *Order and Disorder: the Health Implications of Eating and Drinking in the Nineteenth and Twentieth Centuries*, East Linton, 2000.
ICREFH VI – Hietala, M. and Vahtikari, T. (eds), *The Landscape of Food: the Food Relationship of Town and Country in Modern Times*, Helsinki, 2003.

ICREFH VII – Jacobs, M. and Scholliers, P. (eds), *Eating Out in Europe: Picnics, Gourmet Dining and Snacks Since the Late Eighteenth Century*, Oxford, 2003.

ICREFH VIII – Oddy, D.J. and Petranova, L. (eds), *The Diffusion of Food Culture in Europe from the Late Eighteenth Century to the Present Day*, Prague, 2005.

ICREFH IX – Atkins, P.J., Lummel, P. and Oddy, D.J. (eds), *Food and the City in Europe since 1800*, Aldershot, 2007.

Acknowledgments

ICREFH warmly thanks Dr Virginie Amilien and her SIFO colleagues for organizing such an efficient and interesting symposium. ICREFH also wishes to acknowledge the kind hospitality of Dr Inger Johanne Lyngø, Curator of the Norwegian Museum of Agriculture and Dr Dag K. Andreassen, Curator of the Norwegian Technical Museum. Thanks are also due to Ross Parker for drawing the graphs, to Judith Oddy for her assistance in editing several chapters and checking scientific terminology, and to Katy Low of Ashgate for advice and encouragement.

The authors, editors and publishers would like to thank the following for permission to use copyright material in this book: *Private Eye* (Figure 5.1). Every effort has been made to trace other copyright holders. If, however, there are inadvertent omissions, these can be rectified in any future editions.

Chapter 1

Introduction

Derek J. Oddy and Peter J. Atkins

Obesity in Europe in the Twentieth Century

This book is a first attempt to write the history of obesity in Europe in the twentieth century. Hitherto, this has not been a topic calling for academic research by historians of food consumption, but the present wide-ranging collection of papers establishes a challenge for obesity to be studied as a key economic, social, cultural and physical phenomenon of the modern era. Until now, obesity has been the subject of a multitude of research papers, primarily written by medical scientists, and in the early years of the twenty-first century its extent has attracted categorization as an epidemic. The causes and treatment of obesity have acquired a level of general interest that has made it a topic included in newspapers on an almost daily basis. In the introduction to this collection of papers the editors seek to recognize the importance of obesity in Europe, to identify the chronology of change in body weight in the twentieth century and to establish the obesogenic factors that have contributed to the secular trend in excess bodyweight. Historians of food have laid great stress on the privation and under-consumption of the first half of the twentieth century. It has now become apparent that the focus of investigation must shift to the excessive consumption that has marked recent decades

A Dynamic Bodyweight Model for the Twentieth Century

The incidence of excess bodyweight in European populations shifted its location in the twentieth century. As living standards rose in the late nineteenth century, men's eating patterns − emphasizing the importance of animal foods for their 'strength-giving' qualities − led to a rise in bodyweight with age as energy expenditure declined. By middle age, upper-class and middle-class men were 'stout' if not markedly overweight but were culturally defined as being in 'good health'. For women, to be fat was to be petit bourgeois − upper-class culture favoured the suppression of appetite, whereas working-class women, putting their husbands' and sons' needs before those of their own and their daughters, had insufficient food to eat, which was best evidenced by their stunted growth and the incidence of anaemia. As urbanization became more extensive, anaemia, known as Chlorosis or 'green sickness', was widespread among young female servants, shop assistants and clerical workers by the end of the First World War.

From 1918 there was a marked change in the cultural, social and physical landscape of European society, which followed from the increasing mechanization of industrial production. After World War II, a more sedentary lifestyle developed from the 1960s onwards. Eating patterns changed least and slowest amongst the working classes so that once the severe economic fluctuations of the first half of the twentieth century became a thing of the past, working-class families began to exhibit weight gain as incomes rose and major variations in consumption were prevented by welfare provisions. While slimming fashions and the youth culture had their origins mainly in the interwar years, women's weight increases reflected their rising incomes as greater workforce participation outside the home developed and as their eating patterns increased to match those of men − including, in the last quarter of the twentieth century, a greater consumption of alcohol. In short, men were fatter than women in the nineteenth century and women became fatter than men in the twentieth.

Much of the literature on obesity has addressed the individual behaviour of consumers. Our contention is that the explanation of trends is better sought at the scale of communities, classes, regions and even nations. The factors are cultural, social and economic; they are integral to our changing lifestyles and to structural changes in modern food systems and they are therefore exceptionally difficult for governments to influence. The present book has value in its desire to look at changes in food consumption patterns country by country. Historical publications using this scale of analysis are rare, although the work of Popkin, James and Monda on the recent 'nutrition transition' in China is along similar lines.[1]

Among the factors usually associated with the transition from undernutrition to obesity, four are worth highlighting from the outset. The first is that urbanization had a profound impact upon diets, particularly in countries that had not experienced much city growth or industrialization before 1900. Customary short-distance food supply chains were increasingly stretched, and migrants left behind their traditional fresh food subsistence. On average, less energy was expended in urban jobs than in agricultural labouring, so the nutritional profile was rather different, especially as sedentary employment in the tertiary and quaternary sectors came to dominate advanced economies. A related point about urbanization is that city design in the twentieth century has favoured the growth of obesity. This does not just mean cities designed for the motor car, such as Los Angeles, where walking to work or to the shops would seem to be an odd choice. *All* modern cities to a certain extent have in-built discouragements to exercise. There is now a growing literature on this phenomenon, which refers to the creation of 'obesogenic environments'. These are neighbourhoods where one or more of the following features are present: poor perceptions of safety, especially due to street crime; poor walkability of streets (street connectivity, traffic, pavements, dog mess); poor access to upper floors of high-rise residential buildings; high density of fast-food restaurants; lack of public parks and playing fields; few gyms or sports clubs; local retail grocers emphasizing

---

1   Popkin 1999, Popkin and Gordon-Larsen 2004, Monda et al. 2007, James 2007.

cheap, energy-dense foods in what are sometimes called 'food deserts'; housing with inadequate facilities for the preparation, cooking and storage of perishable foods such as fruit and vegetables; low levels of social networking and support; and, most common, low disposable incomes.[2]

Third, there is an argument that combines demand and supply. On the one hand consumers' standards of living rose steadily in the twentieth century, despite the temporary set backs of wars and economic slumps, to the point where purchasing sufficient food to satisfy hunger was possible for nearly all. On the other hand, the food-supply systems employed to meet this demand were fundamentally transformed. This not only meant a supermarket retail revolution, but also an increasing proportion of processed foods on sale and in the diet and, most significant for obesity, the availability of high-fat foods at affordable prices. The consequence has been a very powerful combination of deeper pockets and a cornucopia of quantity, quality and variety, which together has encouraged over-consumption.[3] Given the difficulty in modern lifestyles of judging a balance between energy intake and exercise, it is hardly surprising that obesity has followed and continues to increase in all of the countries of Europe.

The fourth factor is a matter of knowledge exchange. This is complex and contradictory because modern media are replete with messages advertising fatty junk foods to children but, at the same time, selling slimming products and health foods. There is some helpful research on the history of food advertising and on the evolution of government food policies with regard to healthy diets, but there is less on the reception of such information and its impact upon the act of consumption.

Several recent publications have addressed the history of body weight. Ina Zweiniger-Bargielowska has looked at 'the culture of the abdomen' in the context of British concepts of manliness and racial fitness in the period 1900 to 1939 and Avner Offer attempted a cultural analysis of the rise of obesity in Britain and America since the 1950s.[4] Their emphasis, and in the related literature, is often on individual bodies.[5] Sander Gilman's recent cultural history of obesity is pitched at the societal level but his book is short on historical detail and a comprehensive and conceptually rich treatment in this area is still awaited. Although the present book is a major step towards a road map of histories of the nutritional transition of Europe, there is tremendous potential for further research in this area.

---

2 For more on obesogenic environments, see Holsten 2008, Cummins and Macintyre 2006, Harrington and Elliott 2009, Nelson and Woods 2009, Smith and Cummins 2009, Townshend and Lake 2009.

3 For a comment on increasing portion sizes, see Nielsen and Popkin 2003.

4 Zweiniger-Bargielowska 2005, Offer 2001. See also Chapter 13 of this volume.

5 See also Schwartz 1986, Stearns 1997.

European Food Consumption in the Twentieth Century

For the first time since its symposia began in Münster in 1989, the International Commission for Research into European Food History (ICREFH) has reduced the time span of its discussions to a single century. Earlier symposia assumed that, given the differing stages of economic development in Europe during the period of industrialization and urbanization, changes in the European diet must have occurred over two to two and a half centuries from the mid-eighteenth century to the present day. The Tenth Symposium held in Oslo, in September 2007, faced a more limited time span by the nature of its theme. To focus on the process by which European countries had progressed from a general state of undernutrition to present-day society in which being overweight is quite widespread, and being obese is less uncommon than it was at any previous time, has occurred in a remarkably short period.

It has not been a steady, even progression. Two world wars not only destroyed liberal bourgeois society but also Europe's economic prosperity. From 1918 to 1922 parts of Europe suffered from famines and deprivation that affected many populations and particularly the children in them. The work of Harriet Chick (1875–1977) in Vienna was most notable.[6] Famines also marked the end of the Second World War, from the south-west Netherlands during the winter of 1944–5 in the west to Russia in the east.[7] Recovery from undernutrition was signalled, in food terms, by a surge of compensatory eating – the *fresswelle* – as it was known in German-speaking central Europe. It can also be seen in Western Europe, as in the United Kingdom. Although this recovery became widespread in northern and Western Europe, political control of consumption east of the 'Iron Curtain' meant that progress remained uneven well into the last quarter of the twentieth century. When political control ended, the ensuing 'catch-up' phase, although compressed in time, had consumption targets in the West which were hurriedly emulated during the 1990s. These consumption patterns depended upon a trend which may be summarized as being one in which the domestic production, storage, preservation, preparation and cooking of food was replaced by an extended food chain, in which industrial production of food, its distribution, processing and retailing to the consumer required an increasingly complex number of steps. This limited the sale of raw food materials and replaced them with industrially prepared dishes and whole meals which required only the application of a brief heating stage – generally of only minutes – before being ready to eat. In the postwar years the development of food technology was initially applauded. Magnus Pyke, the eminent food technologist, who wrote the United Kingdom's wartime *Manual of Nutrition*, likened the successes of food science and technology to a 'social

---

6   Weindling 2007.

7   Lindberg 1946 provides a League of Nations' survey of European energy intakes in World War II.

service' entrusted to 'keep the community alive and in tolerable health'.[8] By the mid-twentieth century modern food technology meant processing food in a clean and frequently sterile environment, so that insanitary conditions, particularly with regard to animal products such as meat or milk, seemed to be a thing of the past. The Second World War placed considerable demands on food technology and its underlying science. The technical leadership of the United States meant that besides refrigeration and food-canning, dehydration could limit the cargo space required for foods to cross the North Atlantic. An essential characteristic of the postwar food production system was the increased use of energy at every stage in the process and the intensification of production methods required by the growing demand for food as world population numbers expanded. Although these technical advances were widely applauded, some aspects raised concern. During the late 1950s and 1960s there were rumours that food was being affected by chemicals such as antibiotics and that food additives, designed to enhance flavour and handling qualities, were causing problems. The German and Dutch margarine scare of the late 1950s and early 1960s led to the withdrawal of one brand and the payment of compensation.[9] Other concerns were raised in the 1950s by the possibility that the use of plastic packaging for food might introduce toxins and also that food materials might be contaminated by radioactivity, in particular, the presence of strontium 90. This anxiety was present up to and including the fall out from the Chernobyl nuclear disaster in 1986. The term 'food scare' has been more recently been extended to include nutritional concerns, for instance, general worries about fat content or specific worries such as the presence of trans-fats, and also disease pathogens. Examples of the latter include Bovine Spongiform Encephalopathy (BSE) in cattle and salmonella in eggs – discovered in 1986 and 1988 respectively. Some scholars believe that the modern trend of food scares is so significant that it is representative of what Ulrich Beck has called the 'Risk Society', a new version of modernity in which citizens are fearful and in which trust in scientists and politicians to solve these matters has evaporated.[10]

By the 1990s this industrialized food production system was not only widespread but also highly standardized since large food-processing business firms had attained a size and scale of operations that fitted them to be termed 'multinational' or 'transnational' corporations operating in a global market. This did not mean that by the end of the twentieth century a single cuisine or food culture had become dominant throughout the world or even throughout Europe, but it did mean that some branded products in the food and drinks market had achieved iconic status internationally and that a general cultural influence supporting this type of food market in Europe had developed. The strongest characteristic

---

8   Pyke 1952: vii.

9   *British Food Journal* January 1963: 1, discussing the BBC television programme 'A Suspicion of Poison', broadcast 28 December 1962. Note also the effect of Rachel Carson's *Silent Spring* when it became available in Europe in the mid-1960s.

10   Beck 1986.

of this pressure on consumers was the Americanization of European society. Its influence was strongest in Northern and Western Europe and in industrial and urban social environments, but the cultural determinant of change was the English language: its use as the medium of film and television, its use as the language of advertising, sport, recreation and leisure, its status as the international language at sea and in the air, and its use as the language of commerce and finance. With one or two exceptions, food multinationals originated in the United States, which by its population numbers and the vast extent of its domestic market had faced and overcome the complex problems of standardization and distribution. The multinationals' branded foods and drinks were introduced to Europe by films, television and magazines or by Americans serving as military personnel based in Europe, or forming part of a multitude of international agencies located in major European centres such as London, Paris, Rome or Geneva.

The increasingly refined foods resulting from greater processing which became available during the second half of the twentieth century are easier to digest and, as sold, have less waste. Whilst this is a gain for the consumer, it came as less physical energy was required, as the use of mechanical transport became widespread, as the nature of work became more sedentary and the domestic environment improved as better heating and more labour-saving devices were introduced.

The Concept of Ideal Body Weight

A number of contributors to this volume use terms such as obese or overweight when making reference to body weights. The following note is intended to avoid the repetition which might result should each contributor in turn define the meaning of specialist terms, though some qualification may be necessary where local variations in the usage occur. Despite the concentration on the twentieth century in this book, the concept of an ideal weight arose much earlier in the nineteenth century as a natural progression from the developing interest in statistics. The pre-eminent scholar in the field was the Belgian, Adolphe Quetelet (1796–1874). He published in 1835 a two-volume treatise *Sur l'homme et le développement de ses facultés, ou Essai de physique sociale* [later issued as *A Treatise on Man and the Development of His Faculties, or an Essay on Social Physics*, 1842] from which arose the concept of 'l'homme moyen' or 'average man' characterized by the mean values of measured variables that follow a normal curve of distribution. Quetelet collected data about many variables in the belief that probability influenced human affairs and that it was possible to determine the average physical and intellectual features of a population. His view of the physique of the average man was expressed by the Quetelet Index corresponding to the ratio of the body's weight expressed in kilograms over the square of height expressed in metres i.e. weight/height (kg/m^2). During the second half of the twentieth century this became known as the body mass index (BMI) and was used to express the amount of a person's body fat in place of skin fold measurements. The categories which are usually allocated to BMI are: Underweight = <18.5; Normal weight = 18.5–24.9; Overweight = 25–29.9; Obesity = ≥30.

A slightly simpler anthropometric calculation was suggested by Dr Pierre Paul Broca (1824–80), a French physician interested in anthropology, who attempted to establish an ideal body weight applicable to both men and women. He expressed the ratio of weight to height as being one kilogram for every centimetre of height above one metre. Broca's ideal body weight (IBW) for men was defined as Weight (kg) = Height (centimetres) – 100 centimetres + 10 per cent, and for women Weight (kg) = Height (centimetres) – 100 centimetres + 15 per cent. Broca's index became popular in France and some central European countries and later formed the basis of American life insurance companies' height and weight tables. Using Broca's Index, being overweight was defined as ideal body weight plus 10 per cent.

International conferences from 1957 onwards have tended to favour the Quetelet index, even though his name has almost disappeared in favour of the term BMI.[11] The current European Prospective Investigation into Cancer and Nutrition (EPIC) has suggested that waist circumference, or the waist-to-hip ratio, as indicators of abdominal obesity, may be better predictors of the risk of disease than BMI. Their results of a survey of over 500,000 Europeans in ten countries support the use of waist circumference or waist-to-hip ratio in addition to BMI for the assessment of the risk of death, particularly among persons with a low BMI. Even measurement of the waist alone is significant and may be easier to obtain.[12] It seems that the wealthy fat man, who was somehow symbolic of power in the nineteenth and early twentieth centuries, was a member of a group that suffered from high rates of morbidity and mortality.

Thematic Structure of the Book

The book has been divided into three parts, reflecting the sub-themes discussed at the Symposium:

Part 1: Food Consumption and Consumer Choice

The four chapters in this part contrast the two extremes of Europe in the twentieth century, not only between west and east but also between market economies and state control. At the western extremity, the United Kingdom was the European society most open to the influence of multinationals and Americanization, while in the east, Russia was held back longest from contact with the market-driven products of the international food industry. The other two chapters in this section

---

11   See Chapter 16: 2.

12   Pischon et al. 2008. The EPIC study is based on 23 centres in ten European countries (Denmark, France, Germany, Greece, Italy, the Netherlands, Norway, Spain, Sweden, and the United Kingdom).

represent the experiences of two parts of the former Hapsburg Empire, the Austrian Tyrol and Yugoslavia's Slovenia, similar in climate, terrain and traditional culture yet separated by Communism from 1946 onwards.[13]

In Chapter 2, Josef Nussbaumer and Andreas Exenberger show that the Austrian Tyrol had experienced hunger from early in the twentieth century until during and after World War II. From the 1950s onwards, the predominantly rural Tyrol was transformed into a principal location for the European leisure industry by hotel building, recreational services and a modern transport network designed for motor vehicles. The contrast in Chapter 3, by Tatiana Voronina, is with the effects of the Soviet political system imposed upon Russia from 1917 to 1989. Civil war caused famine, which returned later in the 1930s on an even vaster scale due to the collectivization of agriculture. When war followed in 1941, the terrible damage and loss of life in Russia meant civilians experienced some 40 years of food rationing and, for many, an even longer period of privation. The Union of Soviet Socialist Republics (USSR) offers a unique case-study of how to change a country's traditional cuisine: massive population movements and institutionalized collective feeding produced a 'Soviet cuisine' in the USSR. In Chapter 4, Maja Godina Golija shows that poverty in Slovenia restricted food availability to domestic production and local markets, which delayed the introduction and acceptance of shop-bought foods. Chapter 5, on Britain after World War II, by Derek Oddy offers a marked contrast with the other countries and regions under consideration: Britain was an advanced urban-industrial country with a rapidly changing food market in the twentieth century, and was recovering from ten years of restricted food consumption under the wartime rationing scheme. This account details the restoration of consumer choices and their modification by health pressures late in the twentieth century.

Part 2: Industrial and Commercial Influences on Food Consumption

The transition from own-grown food production to shop-bought food is a key element in European dietary change in the twentieth century. Five contributions in this section show how foods became identified in consumers' minds and how brands and labelling assured recognition by consumers and established quality standards. Brands and trademarks have long been seen by economists, such as Edith Penrose, as essential tools by which the large corporation might grow, and extend control over its market.[14] Chapter 6 by Hans-Jürgen Teuteberg adds to this dimension the pressure of advertising to fix food manufacturers' products in consumers' minds. In Chapter 7, Gloria Sanz Lafuente gives readers an insight into the food market in Spain under the dictatorship prior to General Franco's death in 1975. Ostracized

---

13   Under its leader, Josip Broz Tito, Yugoslavia, while a Communist regime, was not part of the Moscow-led Warsaw Pact. Yugoslavia was a founder member of the non-aligned movement.

14   Penrose 1951, 1959.

by the United Nations in 1945 after World War II, relations gradually thawed in the 1950s as Franco's anti-communist views were seen as an asset in the Cold War. Labelling information became important for Spain to establish prestige for its food products as international economic relations opened during the 1950s and 1960s. Gun Roos, in Chapter 8, discusses the formation of nutritional policy in Norway, with respect to food labelling. Early in the twentieth century opposition from agricultural, trade and industrial interests prevented any government control of foodstuffs; this delayed legislation until the 1930s. Interest in food labelling in Norway is a very recent reflection of health concerns though, despite increasing individual empowerment, passive consumer opinion has yielded more power to food manufacturers. Alain Drouard in Chapter 9 distinguishes France as one country in Europe producing both cane and beet sugar but concentrates more on the beet sugar production, particularly under the Common Market, when marked plant modernization leading to semi-automation of production facilities took place. As the market for sugar in France approached saturation, French producers used research and information supplied by producers' organizations to defend their product against links with weight gain and dental caries and resist the use of artificial sweeteners. The final chapter in Part 2, Chapter 10 by Unni Kjærnes and Runar Døving, starts from the premise that governments can and should control their populations' food choices. Norway's prescriptive moral values produced the *matpakke*, or packed lunch, the Oslo breakfast and, in this chapter, the 'Sigdal breakfast'. In doing so, Norway's social democrat governments separated their people from other Scandinavian countries. However, there were limits to their powers as the authors show in relation to fat and sugar.

Part 3: Social and Medical Influences

The five chapters in this section have remarkably diverse approaches to the question of how social and medical interests have affected food consumption. Jürgen Schmidt seeks, in Chapter 11, to evaluate how nutrition theory was presented in the popular press in Germany and to assess its influence. Here, for the first time, is the modern body image presented as ideal, which gives rise to the contradiction of 'proper' nutrition as seen by the state and the desires of individual. In France, as Julia Csergo shows in Chapter 12, medical discourse in the nineteenth century took a different line from the German-led study of experimental physiology which influenced the USA and Great Britain. French medical opinions frequently sought to explain weight gain by the contribution of non-nutritional factors. When France began to take note of German health policy and observed the connection between nutrition and various illnesses, criticism of industrial catering for the working classes was accompanied in the interwar years by a multitude of diets for the better-off, offered by medical 'experts'.[15] This is matched in Chapter 13 by Ina Zweiniger-Bargielowska identifying a not very different slimming culture which developed in Great Britain at a similar time. Weight reduction was driven by fashion, especially in the case of young women. The slimming movement played into the hands of 'faddists' since Britain lacked an overall nutrition policy. Chapters 14 and 15 deal with central Europe during the Cold War: Czechoslovakia was seen as the 'fattest nation' in Europe by its popular press and in Chapter 14 Martin Franc notes the contrasting demands placed upon its people by the state, its health reformers and fashion images. In fact, in a previous ICREFH volume he has shown that arguments for and against a Czech national cuisine led to contradictions in the state's aims, though then, as in this account, supply difficulties frustrated the authorities' plans.[16] The final study, by Ulrike Thoms, also looks behind the Iron Curtain into East Germany, contrasting its nutrition with that of West Germany from 1945 until reunification. Health policies in the divided Germany reflected the different regimes in the light of the extent and treatment of diabetes.

This book illustrates a major economic and social trend in Europe in the twentieth century which has great significance for the health of its contemporary population. While more research is required, for instance, on countries not included as case studies in this volume, a new sub-theme of European food history has been opened up by this Symposium. Its significance lies in the need for each country to gear its anti-obesity policies to its own specific historical dynamics.

References

Beck, U. *Risikogesellschaft – Auf dem Weg in eine andere Moderne*, 1986, later published as *Risk Society: Towards a New Modernity*, London, 1992.

Carson, R. *Silent Spring*, Boston, 1962.

Cummins, S. and Macintyre, S. 'Food Environments and Obesity: Neighbourhood or Nation?' *International Journal of Epidemiology* 35, 2006, 100–04.

Harrington, D.W. and Elliott, S.J. 'Weighing the Importance of Neighbourhood: a Multilevel Exploration of the Determinants of Overweight and Obesity', *Social Science and Medicine* forthcoming, 2009.

Holsten, J.E. 'Obesity and the Community Food Environment: a Systematic Review', *Public Health Nutrition* 12, 2009, 397–405.

James, W.P.T. 'The Fundamental Drivers of the Obesity Epidemic', *Obesity Reviews* 9, supplement 1, 2007, 6–13.

---

15   As both Chapters 11 and 12 have been written by historians, readers should note that they have set their data in the context of the nutritional knowledge of the day. Contemporary recommended nutrient intakes may have changed significantly from some of those quoted. Recommended protein intakes, for example, underwent marked reductions during the course of the twentieth century.

16   See ICREFH VIII: Chapter 20.

Lindberg, J. *Food, Famine and Relief 1940–46*, Geneva, 1946.

Monda, K., Gordon-Larsen, P., Stevens, S. and Popkin, B.M. 'China's Transition: the Effect of Rapid Urbanization on Adult Occupational Physical Activity', *Social Science & Medicine* 64, 2007, 858–70.

Nelson, N.M. and Woods, C.B. 'Obesogenic Environments: Are Neighbourhood Environments that Limit Physical Activity Obesogenic?' *Health & Place* forthcoming, 2009.

Nielsen, S.J. and Popkin, B.M. 'Patterns and Trends in Food Portion Sizes, 1977−1998', *Journal of the American Medical Association* 289, 2003, 450−53.

Offer, A. 'Body Weight and Self-Control in the United States and Britain since the 1950s', *Social History of Medicine* 14, 2001, 79–106.

Penrose, E.T. *The Economics of the International Patent System*, Baltimore, 1951.

Penrose, E.T. *The Theory of the Growth of the Firm*, New York, 1959.

Pischon, T. et al. 'General and Abdominal Adiposity and Risk of Death in Europe', *New England Journal of Medicine* 359, 2008, 2105−20.

Popkin, B.M. 'Urbanization, Lifestyle Changes and the Nutrition Transition', *World Development* 27, 1999, 1905−16.

Popkin, B.M. and Gordon-Larsen, P. 'The Nutrition Transition: Worldwide Obesity Dynamics and their Determinants', *International Journal of Obesity* 28, 2004, S2–S9.

Pyke, M.A. *Townsman's Food*, London, 1952.

Schwarz, H. *Never Satisfied: a Cultural History of Diets, Fantasies and Fat*, New York, 1986.

Smith, D. and Cummins, S. 'Obese Cities: How Our Environment Shapes Overweight', *Geography Compass* 3, 2009, 518−35.

Stearns, P.N. *Fat History: Bodies and Beauty in the Modern West*, New York, 1997.

Townshend, T. and Lake, A.A. 'Obesogenic Urban Form: Theory, Policy and Practice', *Health & Place* forthcoming, 2009.

Weindling, P. 'From Sentiment to Science: Children's Relief Organizations and the Problem of Malnutrition in Inter-War Europe', *Disasters* 18, 2007, 203−12.

Zweiniger-Bargielowska, I. 'The Culture of the Abdomen: Obesity and Reducing in Britain, circa 1900–1939', *Journal of British Studies* 44, 2005, 239–73.

PART 1
Trends in Food Consumption and Consumer Choice

Chapter 2

Century of Hunger, Century of Plenty:
How Abundance Arrived in Alpine Valleys

Josef Nussbaumer and Andreas Exenberger

The history of the first half of the twentieth century in Europe tends to neglect, amongst other things, the history of hunger. In particular, after the Great War nutrition was insufficient in many regions of the continent. This was true in Russia, as Tatiana Voronina reminds us, and in other places as well, particularly remote areas such as the Alps, where 'hunger' was an official cause of death until the 1920s.[1] Achieving sufficient food was always a serious challenge in 'bad' years. However, during the second half of the century the situation changed significantly: immediately after World War II the average daily energy intake was below 1,500 kilocalories (kcals) per head but increased approximately 2.5 times in the next half century. Consumption patterns changed until nowadays a different form of malnutrition is the most serious problem: overweight is widespread to the point of obesity.

We want to analyse these changes by means of a case study of an Alpine region, the Tyrol in Western Austria, which provides useful insights into the mechanism of how food consumption changed the patterns of demand, local production, and trade.[2] In the case of the Tyrol, the three main transmission mechanisms of change were the immense growth of wealth accompanied by a radical change in the economic structure of the country, the emergence of mass tourism, and the progressive ease of transportation.

Memories are Short-Term: A Nutrition History of the Tyrol

The Dawn of the Century: Agrarian Hungry

At the beginning of the twentieth century the Tyrolean economy was almost completely agrarian. There was emerging industry, particularly along the railways, and in the cities many people were engaged in trade and services, but in the countryside agriculture was dominant and remained the source of political

---

1   Voronina, in this volume.

2   A comparable case study focusing on the shift from subsistence production to market consumption from another part of the Alps is provided by Godina Golija in this volume.

power. Table 2.1 shows the considerable changes in employment during the twentieth century. At the end of the nineteenth century almost three-quarters of the population depended upon agriculture, but, at the beginning of the third millennium, agricultural employment had fallen to a marginal amount that was even below the Austrian national average of 3.9 per cent.

Table 2.1   Employment Share of the Economic Sectors in the Tyrol 1910–2001

Year	Primary (%)	Secondary (%)	Tertiary (%)
1910	56.0	23.0	21.0
1923	55.0	24.0	21.0
1934	49.0	26.0	25.0
1939	46.0	28.0	26.0
1951	36.5	37.0	26.5
1961	24.9	42.9	32.2
1971	10.2	41.0	48.9
1981	6.3	36.0	57.7
1991	4.1	32.8	63.1
2001	2.8	27.3	69.9

Source: Austrian Census, authors' compilation.

Employment in the primary sector did not correspond with the nutritional status of the population since the food supply was always inadequate. One important reason was the natural environment, which is geographically as well as climatically unfavourable for agriculture.[3] One example of the socio-economic effects of this environment were the *Schwabenkinder* ('Swabian children'), who until the 1920s, as young 'surplus eaters', were sent by farmers from Alpine regions during the summer to food-rich Southern Germany (Swabia) to work.[4] But there is also a lot of anecdotal evidence: when for example a hiking tourist (a Mr Zipflhuber) cooked for his companions in a mountain cottage in the evening, the whole pan fell on the ground and the meal was lost. While cooking the meal afresh:

> *Und während der zerknirschte Zipflhuber sich unverdrossen an die Neuarbeit
> machte, näherte sich demütig der Hirte, welcher mit seinen beiden Jungen die*

---

3   Only around ten per cent of the surface area is at all suitable for agriculture. Winters are cold and snowy and growing periods are short and vulnerable to disruptions.

4   See Uhlig 1978. See also Kaser 2008 for a broader picture of child labour in the region.

ganze Scene aus dem Hintergrunde beobachtet hatte, und fragte, ob wir das auf dem Boden schwimmende Zeug [a goulash] *noch zu essen gedächten. Auf unsere lachende Verneinung machten sich die drei darüber her und nach wenigen Minuten gab nur mehr ein großer Fettfleck Kunde von der Katastrophe.* [And while the contrite Zipfhuber assiduously got down to work again, the herdsman [their guide], who had observed the whole scene from the back with his two boys, approached humbly and asked, whether we still intended to eat the mess lying on the ground. After our laughing denial the three of them pitched into it and a few minutes later only a big grease spot told of the catastrophe.][5]

By contrast with today, hunger and extreme food shortage was a recurrent experience during the first half of the century. In the example above, the local tourist guide and his two sons ate what others, unintentionally, had thrown away. While this episode points to widespread endemic undernutrition, near crises in food supply happened cyclically in periods lacking economic stability.

The Ruminants Envied: World War I and its Consequences

World War I was not as short as intended by the Austrian leadership. This particularly hit the Tyrol, which was close to the Italian front line. By May 1918 people even starved, in the city of Trent[6] for example, and the *Innsbrucker Hungerchronik* wrote as early as March 1916 that 'In this famine one could envy the ruminants'.[7] Later, the cleaning of market places was often not by street sweepers, but by people eating rotten food directly from the street. The following quotations from the 'hunger chronicle' give an overview of the events:

11 April 1917	*Ein trauriger Anblick, wenn der Hunger schon so überhand nimmt, dass Weiber sich nicht scheuen, die halbrohen Kuttelflecke auf dem Marktplatz zu verschlingen!*	A sad sight, when hunger gets the upper hand, so that women are no longer shy of gulping down the almost raw pieces of tripe in the market place!
30 November 1917	*Das Brot, das jetzt dem Volke geboten wird, verdient nicht mehr den schönen Namen; es ist eine Mischung, die man früher den Schweinen oder dem Geflügel als Futter vorwarf.*	The bread now given to the public no longer deserves this beautiful name; it is a mixture which in the past would have been given to pigs or fowls.

---

5   The scene from 1899 is quoted in Kramer 1983: 196.
6   Rettenwander 1997: 215. Trent is today the capital of the Italian Trentino region but was part of Habsburg Tyrol until 1918.
7   See Neugebauer 1938. Innsbruck is the capital of the Tyrol.

10 May 1918	*Der Magen ist so schwach geworden, dass er Kartoffel nicht mehr vertragen kann.*	The stomach has become so weak that it can no longer cope with potatoes.
24 July 1918	*Um den Hunger zu stillen, schweifen die Leute durch die Wälder auf der Suche nach Schwämmen und Beeren, schneiden die Ähren von den Halmen und mahlen sie in der Kaffeemühle samt den Grannen und Spelzen.*	To satisfy hunger, people ramble through the forests looking for nuts and berries, cutting the ears of grain from the stalks and grinding them in their coffee mills with husks of barley and spelt wheat.
8 September 1918	*Das Volk fängt an, die Verstorbenen zu beneiden, die dem gegenwärtigen und künftigen Hunger entronnen sind.*	People begin to envy the dead, who have escaped actual and future hunger.

At the beginning of the twentieth century the Tyrol was not self-sufficient in food and the region depended on imports from other Austrian regions as well as from abroad. During World War I, this trade came to a halt. The import of grain and legumes decreased from 16,313 railway cars in 1914 to only 4,244 (−74 per cent) in 1917.[8] Local production suffered as well when horses and oxen were confiscated and conscription reduced the workforce,[9] so that some fields and meadows could no longer be maintained. The rye harvest between 1914 and 1917 dropped from 404,000 cwt to 254,388 (−37 per cent), the potato harvest from 1.73 million cwt to 1.13 (−35 per cent), and with the hay harvest halved within two years, livestock numbers decreased by one-third. Already by 1916 soldiers began to beg civilians for food and in the winter of 1917–18 supplies collapsed, particularly in the Southern part of the country.[10] Even in the well-known spa, Meran, one of the largest and most luxurious cities, bakeries remained out of production for weeks because of flour shortages.[11]

War Goes, Hunger Stays: The Tyrol 1918–24

In November 1918, some hundreds of thousands of soldiers left the Southern front for home.[12] This disorderly mass retreat resulted in the final collapse of supplies,

---

8   Rettenwander 1997: 209.

9   Rettenwander 1997: 60ff.

10   Rettenwander 1997: 196 and 238.

11   Now Merano, Italy. Rettenwander 1997: 215. For a very different vision of Meran with respect to food (or better: drinking), and related to the influence of tourism on it, see Haid 2003.

12   Gasteiger 1986: 271 and 286. The population of Northern Tyrol was doubled by that inflow while the Southern part of the country was annexed by and later awarded to Italy.

especially in the cities. Hans Vonmetz, for example, recalled that in Innsbruck, in November 1918, even a horse, lying in the street and in an advanced state of decay, was hacked at by passers-by and the meat carried away in buckets.[13] By January 1919 food deposits in Innsbruck were exhausted; there were only some hundredweights of flour, very little bread which incorporated ground straw, and almost no milk, fat, potatoes or sugar left.[14] In February 1919, the municipal council stated that '*die Lebensmittelmengen, die die städtische Bevölkerung erhält, [...] gerade ein Drittel von dem [sind], was nach Aussprache der Ärzte zum Leben notwendig ist*' [the volume of food which the urban population receives [... is], according to physicians, just a third of what would be necessary for survival].[15] This allocation was on average only about 800 kcals per head per day.

From 28 April 1919, the *Rucksackverkehr* ('smuggling' of food into the city in backpacks) was tolerated for the following amounts per person: a litre of milk, a kilogram of meat, a kilogram of smoked meat and 12 eggs. Wheat, flour, butter and cheese remained strictly prohibited. On 14 July 1919, a general strike was declared because of insufficient nutrition. On 4 December 1919, hunger riots, the *Innsbrucker Hungerkrawalle*, took place (smaller riots had happened earlier, even during the war, particularly to demand food for children). People gathered in front of the provincial government building to renew the July demands (seeking a guarantee that food would be available against ration cards, more controls on farmers, action against illicit trade). The next day groups of people stormed – among other places – the municipal food depot and a monastery.[16] This escalation provoked international help which finally prevented further riots.[17]

Afterwards the situation improved slowly. Nevertheless it took until December 1921, three years after the war, before the system of ration cards for flour and bread was removed, and the food market returned (in theory) to full freedom of transactions.[18] However, in reality this freedom was restricted by purchasing power which was further aggravated in the following years by war-related hyperinflation. Already in 1920 an inflationary process was observable, labelled the '*schleichender Hungertod*' [creeping starvation] by the local newspaper.[19] Later the situation worsened as shown in Table 2.2: A carpenter's mate earned a weekly wage of 331,152 *Kronen* (kr.) in December 1923, compared to only 29kr. in 1914.[20] While this money bought almost the same amount of bread (61 instead of 63 kg), it bought only 3.7 kg (instead of 6.9) of coffee, or 5.5kg (instead of 9) of butter, and its equivalent in coal declined by more than half.

---

13   Gasteiger 1986: 355.
14   Gasteiger 1986: 410.
15   Quoted in Gasteiger 1986: 412.
16   Gasteiger 1986: 363ff.
17   Gasteiger 1986: 372f.
18   Rettenwander 1997: 361.
19   *Innsbrucker Nachrichten* March 23, 1920: 1.
20   ÖSZ 1924: 93.

Table 2.2   Prices of Selected Goods 1914–23 in Austrian Krones

	Prices				Annual Average Inflation	
	Jul 1914 (kr.)	Dec 1921 (kr.)	Dec 1922 (kr.)	Dec 1923 (kr.)	1921-23 (%)	1914-23 (%)
1 kg potatoes	0.16	55	750	1,800	472	167
1 kg bread	0.46	54	5,200	5,400	900	168
1 kg flour	0.52	580	6,800	6,600	237	170
1 kg sugar	0.84	300	8,800	11,400	516	172
1 egg	0.08	100	2,500	2,375	387	196
1 kg butter	3.20	2,400	50,000	60,000	400	182
1 kg lard	1.86	2,200	27,000	31,000	275	178
1 kg meat	1.70	800	14,000	30,000	512	180
1 kg pork	1.96	1,600	22,000	36,000	374	181
1 kg coffee	4.20	3,000	63,000	90,000	448	186
1 kg wood	0.03	7	440	440	693	175
1 litre petroleum	0.38	187	3,200	3,500	333	161

Sources: ÖSZ, 1923, 82ff. and ÖSZ, 1924, 89ff. (inflation rates are authors' calculations).

The Great Depression

Hyperinflation and its consequences had not been forgotten when, from 1929, the Great Depression affected the region, doubling unemployment to almost one-third of the non-agrarian workforce until 1933.[21] Consequently, about 50,000 people, or 14 per cent of the population, were officially in a state of famine in 1933, particularly those without any entitlement to support.

In 1932, a food survey among Tyrolean textile workers was conducted by the local workers' organization. The result was alarming: only 18 per cent had meat more than once a week, 40 per cent once, on Sundays, and 42 per cent recorded meat only at festivals.[22] Another survey of 120 families in 1933 came up with even worse results: 52 households (43 per cent) used no type of meat at all but which, if it was mentioned, was usually horse meat, 105 (88 per cent) no butter, and 103 (86 per cent) no coffee.[23] The daily average energy intake per head was only 2,215 kcals, but only 1,724 kcals in families with more than five people. This was 15 per cent less than in other regions of Austria and considerably below

---

21   Kleon-Praxmarer and Alexander 1994: 239.
22   Kleon-Praxmarer 1990: 90 (from *Arbeiterkammer Innsbruck*).
23   Kleon-Praxmarer 1990: 86ff. See also Nussbaumer 1980: 156f.

the level of 2,400 kcals regarded as the minimum for normal consumption in the 1920s by the League of Nations.

The Penultimate Hunger Crisis in the Tyrol

After the Great Depression the situation improved only slightly, but the worst was over by the middle of the 1930s. When the Nazis took over the Tyrol in March 1938, their priority, in preparation for war, was that the food shortage during World War I must not be repeated. Hence, the free market was replaced by a rationing system. Ration cards remained convertible at least until the beginning of 1943. In the Tyrol, from 1939 to 1944, the potato harvest declined only a little from 58,000 tons to 51,000, while milk production even increased slightly from 174,000 tons over the same period. To achieve sufficient supplies, domestic farmers were tightly controlled. Prisoners of war and *Fremdarbeiter* (forced labourers from occupied territories) filled the vacancies in the work force, and imports from occupied territories secured full food provision. This policy resulted at the same time in starvation in other regions controlled by German forces.[24]

Nevertheless problems began in March 1942, when newspapers announced the reduction of food rations.[25] While daily rations represented 2,600 kcals per head in 1939, they were reduced to 1,700 kcals until 1943 and to 1,500 kcals until February 1945, when the regime had overstretched all its resources to save public support from collapsing.[26] Nevertheless all efforts failed, and even the *Sicherheitsdienst* [Security Service] in the Tyrol had to record critical voices:

> *Wenn ich nichts mehr zum Essen bekomme, gehe ich auch nicht arbeiten. Wenn sie mich einsperren, habe ich wenigstens das Essen und auf die Kinder werden sie noch schauen müssen! Uns Alten geben's immer weniger zum Fressen, wir sollen krepieren, damit sie nicht so viele Leute zum Versorgen haben. Die Hauptsache war, dass wir ihnen die Kinder großgezogen haben.* [If I do not get something to eat, I may as well not go to work. If they imprison me I will at least have food and they will have to care for my children at last. Us elderly are given less and less to eat; we will croak so that they have less people to care for. The main thing was to raise the children.][27]

---

24 See Nussbaumer 2003: 185ff.

25 Schreiber 1994: 109 and 188f. Furthermore, flour was already everything but pure, and cigarettes were of particular use to suppress hunger feelings.

26 Schreiber 1994: 191.

27 Quoted in Schreiber 1994: 188f.

The Disastrous Eve of Hunger in the Tyrol

In May 1945, Europe was in ruins. Food supplies were halved even from the extremely low levels of February until the summer and food became the most urgent issue. The local newspaper's headlines in August 1945 were: '*Was wir jeden Tag essen, womit wir kochen sollen, das ist heute unsere Hauptfrage*' [What we eat every day, what to cook, is our main question today], while Colonel Sazerac, director of nutrition in the French-occupied zone, announced in May 1946 that domestic food production in the French zone could only provide 440 kcal per head per day.[28] For that reason people used all available resources to produce food: keeping chickens on balconies became common as well as using public lawns and gardens in the city to grow vegetables. Figure 2.1 shows that official rations were completely insufficient, a state which continued until 1948. In the summer of 1947 these numbers translated into an allocation for four weeks per head of 6 kilograms (kg) of bread or 200 grams (g) per day, 1 kg maize, 500 g fat, 400 g meat, 300 g salt, 250 g each of flour, pasta and sugar, 150 g soup cubes and 120 g cheese.[29]

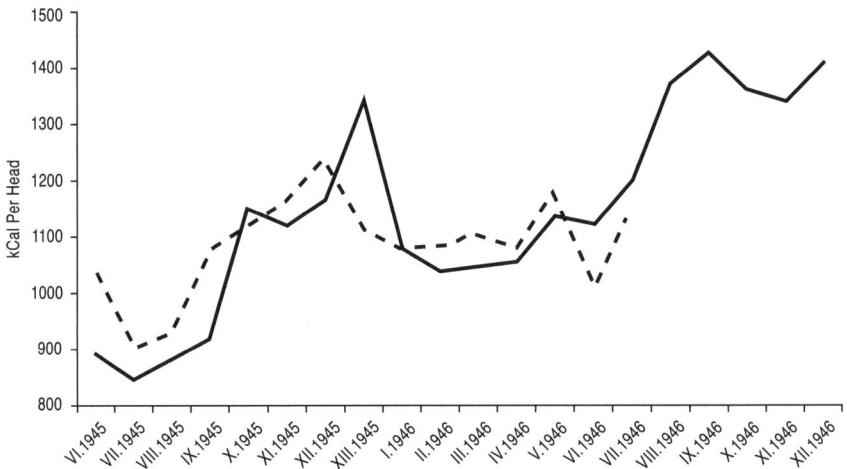

Figure 2.1 Food Rations for Adults in Innsbruck 1945–46 (kcal/head/day)

Sources: Nussbaumer, 1992, 32; Eisterer, 1986, 196.

Notes: Data are reported in four-weeks-periods counted in roman numerals, hence there are 13 per year; the series shown starts on 29 May, 1945 (beginning of VI.1945), and ends on 8 December, 1946 (end of XII.1946); the solid line shows data from the *Statistical Yearbook of Innsbruck* and the *Annual Report of the Tyrolean Workers' Organization*, the dotted line shows alternative data from the French occupational forces.

---

28   *Tiroler Tageszeitung* August 7, 1945: 1.
29   Nussbaumer and Exenberger 2006: 125.

Only in the summer of 1948 did rations begin to provide more than 2,000 kcals per head per day.

Finally, after four years of suffering, the worst, in the sense of extreme hunger, was over. Although even in the early 1950s some shortages of milk, fat and meat still occurred, rationing of bread, pasta, cheese and fish was removed in 1949 and for all other goods during the early 1950s.[30] The people could hardly believe that hunger would not return.

Postwar Recovery: The 'Wirtschaftswunder'

The growth of wealth after Word War II was perceived as miraculous and hence labelled the *Wirtschaftswunder* [economic miracle]. Austria, like neighbouring West Germany, was subject to an economic boom without precedence in European history. The average growth of Gross Domestic Product (GDP) per head was 5.8 per cent in the 1950s (compared to 2.8 per cent world wide) and remained at 4.1 per cent (compared to the world figure of 3 per cent) in the 1960s.[31] This is at the heart of what became known as the '1950s-syndrome' through which the new wealth was at first mainly converted into better nutrition – quantitatively to begin with (in the *Fresswelle*, a virtual 'eating wave') – followed by improvements in quality and only later into other improvements of living standards.[32]

Several factors transformed the Tyrolean economy. From the middle of the twentieth century the road network was built up, including motorways parallel to the nineteenth-century railways. Two indicators of that are the growth of road mileage and the intensity of traffic across the Brenner Pass, the main north-south transit route through the Tyrol.

---

30   Nussbaumer 1992: 39ff.
31   Maddison 2003. Accurate regional data for the Tyrol is missing.
32   Bandhauer-Schöffmann and Hornung 1995.

Table 2.3   Length of Certain Types of Road in Austria, 1948–2000

	Motorways (km)	Federal (km)	Provincial (km)	Communal (km)
1948	0	694	1,174	–
1964	6	1,021	1,106	5,883
1984	141	1,173	1,228	ca. 5,900
2000	216	1,002	1,270	–

Sources: Nussbaumer, 1992, 151; IVT, 2001, 2.

Notes: 'Motorways' include various types of multilane roads; 'federal' and 'provincial' roads are both regional connecting roads.

Table 2.4   Transportation across the Brenner Pass 1960–2000

	Railways (M tons)	Road (M tons)	Total (M tons)
1960	2.7	0.4	3.1
1970	3.6	3.1	6.7
1980	4.3	11.2	15.5
1990	6.5	13.6	20.1
2000	10.0	28.0	38.0

Source: Nussbaumer and Exenberger, 2006, 133.

While Table 2.3 shows the extension of the road and motorway system, Table 2.4 indicates the intensity of through traffic. A further indicator of the significance of transportation and also for the growing ease of supply over longer distances was the number of passenger cars and freight vehicles. Both grew steadily over the decades following World War II. In 2000 there were 28,741 freight vehicles and 329,911 passenger cars (almost 0.5 per head) in the Tyrol, which points to an enormous increase in the possibilities for buying, selling and transporting goods.

Tourism was a major reason for Austria's prosperity, and the Tyrol has become one of the most concentrated tourist regions in the world with tourist numbers usually exceeding 8 million per year. This had four different effects on the local food market:

- producers profited due to the tourists' purchasing power, which increased the demand for local products;
- tourists' demand changed food habits, causing further food imports and hence problems for local producers;

- consumers had more choice, but had to pay higher prices for food, particularly during the tourist seasons; but
- producers profited because tourists discovered Tyrolean food specialities and created a demand for them in foreign markets.

The last point became increasingly important after Austria's accession to the European Union in 1995, when European markets were opened for Austrian products.

The growth in tourist numbers took place mainly up to the early 1980s. A growth in the quality of holidays followed and also a structural change, when winter tourism overtook summer tourism in the 1990s. In 2006, tourists accounted for 41.7 million overnight stays, which meant an extra population of 114,000 (16.4 per cent) on average in that year. However, tourists tend to cluster geographically as well as temporally. If the country is 'fully booked', as happens during peak winter holidays, tourists increase the overall population by 49 per cent and almost one-third of all communities possess more tourist beds than inhabitants. In some cases the potential 'extra' population exceeds the resident population by two, three or even five times, as Table 2.5 shows. Infrastructure and food supplies are necessary for this extra population, and their provision is particularly costly and difficult in winter and in remote valleys, where most of the leading resorts are located.

Table 2.5   Tourism in the Tyrol 1950–2000

	Millions of overnight stays			'Extra' population	
	Winter	**Summer**	**Total**	**000s**	**per cent**
1950	0.6	1.7	2.3	6.4	1.5
1960	2.8	8.9	11.7	31.9	6.9
1970	7.8	17.9	25.7	70.4	12.9
1980	15.9	22.9	38.7	105.8	18.0
1990	20.6	21.9	42.5	116.3	18.4
2000	22.4	17.4	39.8	108.8	16.1

Source: Nussbaumer and Exenberger, 2006, 135 (with assistance of Paul Tschurtschenthaler), data from the Tyrolean Census Bureau.

Note: 'Extra' population is calculated by dividing total overnight stays by the number of days, in relation to the census population.

In a city like Innsbruck the population is not significantly increased by tourists, but some smaller communities are actually tourist villages with 'native' shares of less than 30 per cent. Popular winter resorts more than double their populations.

And, with respect to built-up areas, the leading tourist destinations accommodate more than 500 people per hectare.[33]

The Impact of the 1950s-Syndrome on Nutritional Habits

By the 1960s, as Figure 2.2 shows, Austrian consumption figures for daily energy intakes (Tyrolean numbers would be slightly lower) were in excess of world trends at around 3,250 kcals per day.

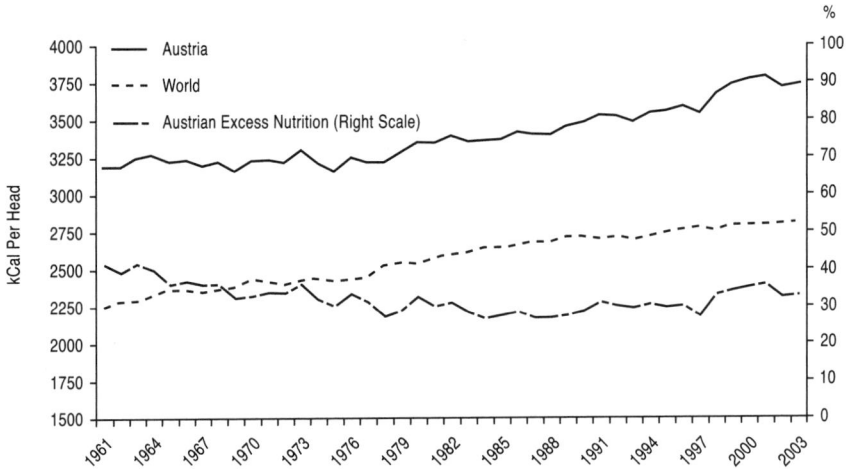

Figure 2.2 Average Food Energy Consumption 1961–2004 (kcal/head/day)

Source: Food and Agricultural Organization, FAOSTAT.

Notes: According to FAO methodology, the series shown actually refers to production, i.e. calories are calculated based on domestic production (for food use) minus exports plus imports, which is only a proxy for actual consumption; 'excess nutrition' is calculated as the Austrian average over the World average minus 1.

The radical change in wealth and consumption patterns – labelled the '1950s-syndrome'[34] – has overturned historical memories. At the end of the 1940s, potatoes (10.5 kg per head per month) and bread (9.6 kg) dominated consumption. As Figure 2.3 shows, two decades later, meat consumption had more than doubled, while the consumption of bread had dropped by 40 per cent and of potatoes by 58 per cent. During the 1970s, meat consumption climbed to more than 4 kg and even

---

33   YEAN 2005: 218f.
34   See Pfister 1995.

surpassed bread as well as potatoes.[35] Another example is eggs, which were still extremely scarce in 1948, but increased strongly in consumption in the following decades, while the price remained the same.[36]

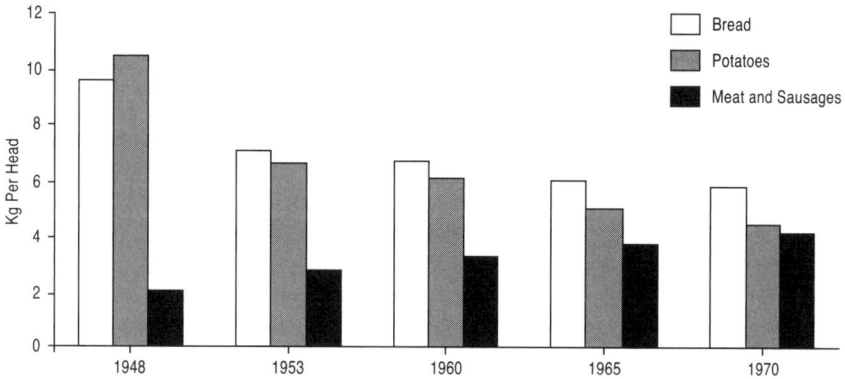

Figure 2.3 Average Tyrolean Consumption of Selected Goods 1948–70 (kg/head/month)

Source: Nussbaumer, 1992, 182; Tyrolean Workers' Organization.

The Growth of Food Surpluses

In close connection with these trends, food 'autarky' collapsed completely in the Tyrol. Production shifted from subsistence needs towards specialization. In the 1980s the region was self-sufficient only in certain cattle products, notably milk and veal, but dependent on imports particularly of grain, vegetable oil, vegetables and fruit. Some of these deficits could be compensated for by intra-Austrian trade, some demanded international trade, as Figure 2.4 shows. Nevertheless, there is of course also the corollary of growing exports. To name just three examples of successful Tyrolean food producers: the biggest milk-producer, Tirol Milch, has recently been exporting about a third of its production; the biggest jam producer, Darbo, more than a third, and the bacon 'tycoon', Handl Speck, more than half.

---

35   Nussbaumer 1992: 182f.
36   Nussbaumer 1992: 42 and 182.

%

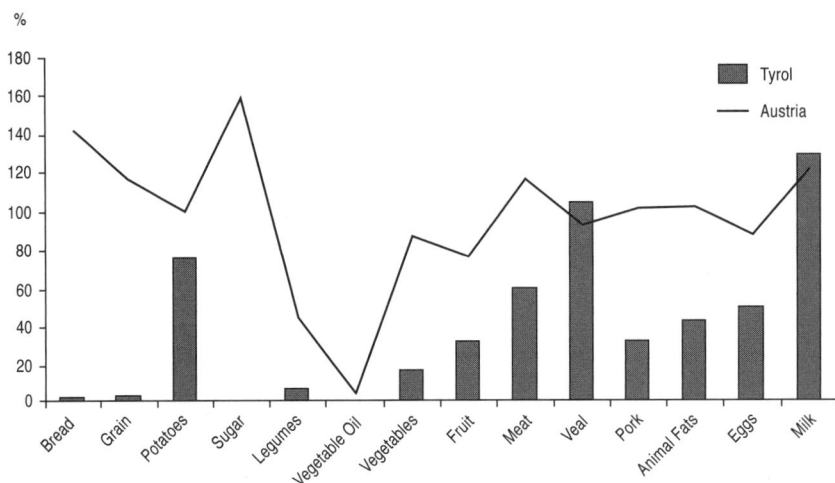

Figure 2.4 Self-Sufficiency in Foodstuffs, Austria and the Tyrol, 1984 (per cent)

Source: Nussbaumer, 1992, 118, from *Landeslandwirtschaftskammer Tirol*.

From Sargol to Xenical

As the whole nutrition pattern changed during the twentieth century, so did the major health problem from too little to too much weight. The pharmaceutical industry profited from both situations. Around 1900 medicines like 'Sargol' were popular, dedicated to overcome deficits in weight by advertising 'Sargol makes you opulent and attractive', while a century later the tremendously successful anti-fat pill 'Xenical' promised 'less weight, more life'. When it was sold in Tyrolean drug stores for the first time in September 1998, supplies for the whole year were sold out within a day.[37]

In the early 1950s it was still being reported that people with a weight deficit of 20 to 30 kgs due to malnutrition were not uncommon.[38] Now the same numbers of people are overweight by the same amount. Today, 6.8 per cent of the Tyrolean population suffer from obesity, the relative increase within a decade being above the national average (though the absolute number is not).[39] Generally, eating disorders are far from negligible as indicated by Table 2.6.

---

37   *Tiroler Tageszeitung* September 2, 1998: 1.
38   Nussbaumer 1992: 211.
39   Klimont et al 2007: 178; Statistik Austria 2007.

Table 2.6   Distribution of Weight Classes and Frequency of Eating Disorders in the Tyrol 1997

	Women (%)	Men (%)
Underweight (BMI < 19)	11.6	4.3
Normal weight (BMI 19–25)	51.6	59.4
Overweight (BMI 25–30)	32.1	31.5
Obese (BMI > 30)	4.8	4.8
Anorexia nervosa (acute)	0.3	–
Anorexia nervosa (biographical)	1.5	–
BE (binge eating)	12.2	–
BES (binge eating syndrome)	8.4	4.2
BED (binge eating disorder)	3.3	0.8
Bulimia nervosa	1.5	0.5
Non-specific eating disorder	–	9.4

Sources: Kinzl et al., 1998a ; Kinzl et al., 1998b.

Notes: Data from 1,000 telephone interviews (two stage random, representative) among women and men each in the Tyrol; the definition of BES is different between men ('binge eating disorder, partial picture') and women ('binge eating syndrome').

Conclusions

At the beginning of the century, the Tyrol was an agrarian society, although never self-sufficient in food. Hence, hunger was prevalent during the first half of the century, particularly during five serious man-made food shortage periods. In the 1950s consumption patterns changed in a sustained manner, leading to quantitative and qualitative improvements in nutrition. Consequently, over-nutrition has become an issue of public health, and excess weight is now as widespread as deficits were earlier. The three most important reasons for these developments are the improvements in transportation and economic growth (which were general trends for Europe) and the manifold influence of mass tourism (which is more specific for the Tyrol). The biggest domestic supermarket chain, M-Preis, assessed the effects of this economic development:

> Society has changed radically in the last 30 years. In former times, when a
> nearly car-free society went shopping primarily on foot, 'proximity' was defined
> completely differently from today. By increased individual mobility the meaning

of this term has shifted: more than 80 per cent of shopping is [now] done by car.[40]

This was reinforced by the continuous increase in road quality, particularly in the form of bypasses for most villages. Consequently, there are already a number of municipalities which lack shops providing basic supplies, sometimes even completely lacking shops. On the other hand, supermarket chains cluster in populated areas and there are also eight shopping centres which, though small by international standards, tend to be located along bypasses, together with fast food restaurants.[41]

As a very sensitive region with respect to nutritional vulnerability for economic, and also geographical and climatic reasons, the Tyrol is a kind of seismograph, revealing general European developments in food consumption and the factors which influence them. It is indicative of general trends as well as deviations from those trends and, for that reason, it is certainly an interesting case for European food history, inviting further research.

References

Bandhauer-Schöffmann, Irene and Hornung, Ela, 'Von der Erbsenwurst zum Hawaischnitzel. Geschlechtsspezifische Auswirkungen von Hungerkrise und "Fresswelle"' [From Hawaiian Schnitzel to Pea Sausage: Gender-Specific Effects of the Hunger Crisis and Eating Wave], in Albrich, T., Eisterer, K., Gehler, M. and Steininger, R. (eds), *Österreich in den Fünfzigern*, Innsbruck, 1995, 11–34.

Eisterer, K. 'Hunger und Ernährungsprobleme in Tirol aus der Sicht der französischen Besatzungsmacht' [Hunger and Nutrition Problems in the Tyrol from the Perspective of the French Occupying Power], in Pelinka, A. and Steininger, R. (eds), *Österreich und die Sieger*, Vienna, 1986, 189–204.

Gasteiger, E. *Innsbruck 1918–1929: Politische Geschichte* [Innsbruck 1918–1929: Political History], 2 vols, Innsbruck University, unpublished thesis, 1986.

Godina Golija, M. 'Food Culture of the Less Affluent Slovene Urban Population and Efforts for its Improvement: The Case of Maribor Municipality, 1900–1940', in ICREFH V, 145–54.

Haid, O. 'Early Tourism and Public Drinking: the Development of a Beer-Drinking Culture in a Traditional Wine-Producing Area', in ICREFH VII, 105–24.

IVT, *Tiroler Verkehr in Zahlen 2001* [Tyrolean Traffic Figures] Innsbruck, 2001.

---

40   Source: http://www.mpreis.at/mpreis.htm [accessed: 1 September 2007, no longer available (translation A.E.)].

41   YEAN 2005: 196f.

Kaser, J. *Das Kapital des armen Mannes. Kinderarbeit im Tirol des 19. Jahrhunderts* [The Capital of the Poor Man: Child Labour in the Tyrol in the 19th Century], *Geschichte & Ökonomie* 19, Innsbruck, 2008.

Kinzl, J.F., Traweger, C., Trefalt, E. and Biebl, W. 'Eßstörungen bei Frauen: Eine Repräsentativerhebung' [Eating Disorders in Women: a Representative Survey], *European Journal of Nutrition* 37, 1, 1998a, 23–30.

Kinzl, J.F., Traweger, C., Trefalt, E., Mangweth, B. and Biebl, W. 'Eßstörungen bei Männern: Eine Repräsentativerhebung' [Eating Disorders in Men: a Representative Survey], *European Journal of Nutrition* 37, 1, 1998b, 336–42.

Kleon-Praxmarer, R. *Die Tiroler Arbeiterschaft in der Weltwirtschaftskrise* [The Tirolean Labour Force in the Great Depression], Innsbruck University, unpublished thesis, 1990.

Kleon-Praxmarer, R. and Alexander, H. 'Tirol Wirtschaft vom Beginn des Ersten bis zum Ende des Zweiten Weltkrieges' [Tyrol Economy from the Beginning of the First to the End of the Second World War], in GFW (ed.), *Chronik der Tiroler Wirtschaft mit Sonderteil Südtirol*, Vienna, 1994, 209–62.

Klimont, J., Kytir, J. and Leitner, B. *Österreichische Gesundheitsbefragung* 2006/07 [Austrian Health Survey], Vienna, 2007.

Kramer, D. *Der sanfte Tourismus: umwelt- und sozialverträglicher Tourismus in den Alpen* [Gentle Tourism: Environmentally and Socially Sound Tourism in the Alps], Vienna, 1983.

Maddison, A. *The World Economy, Historical Statistics* (CD-rom), Paris, 2003.

Neugebauer, H. 'Innsbrucker Hungerchronik 1915–1918' [Innsbruck Hunger Chronicle], *Tiroler Heimatblätter* 16, 1938, 261.

Nussbaumer, J. *Sozial- und wirtschaftsgeschichtliche Aspekte der Weltwirtschaftskrise in Österreich 1929–1934* [Socio-Economic and Historical Aspects of the Great Depression in Austria], Innsbruck University, unpublished thesis, 1980.

Nussbaumer, J. *Sozial- und Wirtschaftsgeschichte von Tirol 1945–1985. Ausgewählte Aspekte*, [Social and Economic History of Tyrol 1945–1985: Selected Aspects], Innsbruck, 1992.

Nussbaumer, J. *Vergessene Zeiten in Tirol. Lesebuch zur Hungergeschichte einer europäischen Region* [Forgotten Times in the Tyrol: Reader on the Story of a Hungry European Region], Innsbruck, 2000.

Nussbaumer, J. *Gewalt.Macht.Hunger 1. Schwere Hungerkatastrophen seit 1845* [Severe Hunger Crises since 1845], Innsbruck, 2003.

Nussbaumer, J. and Exenberger, A. 'Per un Quadro Generale dell'Economia Tirolese nei Primi Anni del Secondo Doppoguerra (con Alcune Implicazioni)' [A Survey of the Tyrolean Economy in the First Years after World War II (with some Implications)], in Bonoldi, A. and Leonardi, A. (eds), *La Rinascita Economica dell'Europa. Il Piano Marshall e l'Area Alpine* [The Economic Reconstruction of Europe: the Marshall Plan in the Alpine area], Milano, 2006, 121–41.

ÖSZ (ed.), *Statistisches Handbuch für die Republik Österreich* [Statistical Handbook for the Republic of Austria] Vol. III, Wien, 1923.

ÖSZ (ed.), *Statistisches Handbuch für die Republik Österreich*, Vol. IV, Wien, 1924.

Pfister, C. (ed.), *Das 1950er Syndrom. Der Weg in die Konsumgesellschaft* [The 1950s Syndrome: the Path to the Consumer Society], Bern, 1995.

Rettenwander, M. *Stilles Heldentum? Wirtschaft s- und Sozialgeschichte Tirols im Ersten Weltkrieg* [Quiet Heroism? Tyrolean Economic and Social History in the First World War], Innsbruck, 1997.

Schreiber, H. *Wirtschafts- und Sozialgeschichte der Nazizeit in Tirol* [Economic and Social History of the Nazi Era in the Tyrol], Innsbruck, 1994.

Statistik Austria 2007, 'Verteilung des Body-Mass Index (BMI, WHO neu) in der Bevölkerung ab 20 Jahren 1999' [Distribution of Body Mass Index in the Population Aged 20]. Available at: http://www.statistik.at/web_de/statistiken/ gesundheit/gesundheitsdeterminanten/bmi_body_mass_index/index.htm [accessed 1 July 2007].

Uhlig, O. *Die Schwabenkinder aus Tirol und Vorarlberg* [The Schwabian Children from the Tyrol and Vorarlberg], Innsbruck, 1978.

YEAN, *TirolCITY: New Urbanity in the Alps*, Vienna, 2005.

Other Sources (Data and Primary Material)

Arbeiterkammer Tirol (provincial workers' organization)
FAOSTAT (Food and Agricultural Organization, statistical data base)
Innsbrucker Nachrichten (local daily newspaper, pre-World War II)
Landesarchiv Tirol (provincial archive)
Landeslandwirtschaftskammer Tirol (provincial farmers' organization)
Landesstatistik Tirol (provincial census bureau)
Statistisches Handbuch der Stadt Innsbruck (local statistical yearbook)
Tiroler Tageszeitung (local daily newspaper, post-World War II)

Chapter 3

From Soviet Cuisine to Kremlin Diet: Changes in Consumption and Lifestyle in Twentieth-Century Russia

Tatiana Voronina

Introduction

The everyday life and meals of the Russian people in the twentieth century were influenced by numerous political, economic, and cultural events. At the beginning of the century food was traditional and retained its essential ethnic features. Political reforms after 1917 initiated a gradual change in daily meals but there was a major shock in the 1920s when Soviet Russia suffered from the Great Hunger and millions of people died. The Second World War (1941–5) brought many troubles, including lack of food, but there was another shock when, in 1946–7, people starved again.

During the Soviet period the ethnic composition of the country changed significantly because of extensive migration between the central Russian Federation and the Union Republics. From an historical perspective, migration is not a recent phenomenon in Russia. As a rule, those who migrated adhered to their traditional diets. As a result, many dishes and foods of other republics became familiar to many Russians and appeared in cookery books, including the one which has remained the most popular since 1939 – *The Book About Tasty and Healthy Food*. In effect, therefore, the so-called 'Soviet cuisine' was an amalgam of various regional dishes and reflected the norms of the Soviet lifestyle. The disintegration of the former Union of Soviet Socialist Republics (USSR) in 1991 resulted in the forced migration of ethnic Russians from the Soviet Republics, who became, in effect, immigrants in the new Russian State. Against a background of the collapse of the home food industry and a drop in living standards for most of the population, many new products were imported from abroad. This resulted in a slow but difficult adaptation to totally unfamiliar foods and preparation methods.

At the end of the twentieth century when the Russian economy had stabilized there was plenty of food to choose from in the shops, but there was a great difference in purchasing power between poor and rich people. A high standard of living resulted in numbers of overweight people who tried to follow various slimming diets, of which the 'Kremlin diet' became the most popular among them thanks to its rapid results. The main purpose of this chapter is to demonstrate the close relationship between political and other factors and consumer choice or 'the cost of the plate'.

Russia lacks academic studies in ethnological science. The source material for this chapter is based upon articles, official communications and advertisements from newspapers, magazines and other periodicals, and everyday broadcasts from various regions of Russia. These sources reveal the socioeconomic situation in the country and the effects on consumer choice. Relevant findings by nutritionists have also been incorporated.

The Peculiarities of Soviet Cuisine, 1917–91

The October Revolution of 1917 initiated a period of considerable change in Russian traditional cuisine.[1] To resolve the problems of food consumption in the spring of 1918, a 'food dictatorship' was adopted – a system of extraordinary measures to supply the army and workers. It meant that all the grain, bread and cattle were taken by the state. In the years 1919–21 this was followed by '*prodrazverstka*' (surplus-appropriation system) – a new system of State grain purchase in which free trade was banned. The main aim of these and other measures (called 'war communism') was to ensure the success of the new government in the Civil War (1918–21). The result was an economic crisis, and in 1921 there was a Great Hunger when millions people died, especially in Volga region.[2] To save the situation a 'New Economic Policy' (NEP) that allowed free trade was introduced as a stop-gap measure. This came to be known as 'state capitalism' with the aim of industrialization as the basis of a new communist society.

The most fundamental changes took place during 'collectivization' in the 1920s with the abolition of individual peasant holdings – and their associated way of life – in favour of collective farms (*kolkhoz*). As a result, a centuries-old way of life was broken. Many traditional cereal crops disappeared as other crops began to be more extensively cultivated. From 1929, collectivization was carried out by the forced liquidation of millions of rich peasants and by the creation of the GULAGs (concentration camps). In 1932–3 there was an extensive famine in which people were said to have 'attacked' the fields, trying to collect ears of grain, but government bodies took punitive measures. A special Act concerning the protection of state and public property was passed on August 7, 1932, known by the shorthand 'seven-eight' or the Act 'about ears'. As a result, almost seven million peasants were subjected to repression and sent to Siberia. At the cost of many lives, a socialist form of agriculture was set up and the USSR became an 'industrial-*kolkhoz* state'.

---

1   Voronina 1998: 246.

2   See the concluding chapter in this volume. Among those who really helped the famine regions was Alfred Nobel.

It is impossible to comment on nutrition without taking into account all the circumstances that led to the Union of the 15 Soviet Republics. Together with the economic changes, there was a large-scale migration from the Republics accompanied by a gradual stream of people from villages to the central part of Russia, and especially to Moscow. Traditional Russian cuisine was augmented by provincial culinary and non-Russian dishes. For example, people from Siberia and the Urals brought *pelmeni* (Siberian meat dumplings) and pies like *shanezhki*, and Belorussians and Ukrainians brought salted fat, noodle-soup, and beef Stroganoff. Cutlets from the Petersburg Merchants' Club came to the Ukraine in 1918 and, a new dish 'cutlet à la Kiev' appeared on Soviet restaurant menus. Cottage cheese pancakes (curd fritters) came from the Baltic States and *borshch* and *vareniki* (curd or fruit dumpling). Tea was a luxury before 1917 and was drunk as a special dessert or at the evening meal, but during the 1920s it became an article of mass consumption and was drunk throughout the day.[3]

Between 1921 and 1931 further socioeconomic developments led to the spread of private dining-rooms and state public-catering establishments. The range of dishes widened and the demand for improved taste and quality rose.[4] In the 1930s, collectivization was accompanied by the broadening of public catering and the building of large-scale mechanized canteens offering limited menus of simple dishes. Spices and flavourings were excluded and boiling, the main form of cooking in the old Russian cuisine, dominated. Thus, in spite of radical changes and novelties, its basis remained unshakeable. *Shchee* (soup with cabbage, potatoes and meat), boiled buckwheat with butter, cranberry *kissel* (a kind of starchy jelly) and tea with jam or lemon became the standard items of Soviet cuisine from the 1920s to the 1940s, for the Kremlin élite and factory workers alike.

Festive food included home-made *pelmeni*, goose or duck with sour apples, noodle-soup with chicken and, of course, lots of *pirogi* – pies with mushrooms, eggs, rice and *viziga* (dried spinal chord of cartilaginous fish). River fish and also smoked fish *balyk* (cured fillet of sturgeon), soft and pressed caviar (black and red) and salted salmon were in abundance. On the whole, domestic menus kept their national features, especially in the east of the country and in the Caucasus, and also in big families of several generations. At the same time, the menu was frugal, meeting the ideals and requirements of the period.[5]

The Second World War (1941–5) brought many severe trials to ordinary life. A rationing system was introduced with a fixed allowance of bread at 125 grams a day. The best cooks were called up for military service to provide food with hot *shchee* and porridge as the main dishes. The ordeals of the war years adversely affected all the spheres of life. It took many years to reconstruct the national economy and the food during which many old dishes and their names and methods of preparation, were forgotten.

---

3   Pohljobkin 1997: 47–8.
4   Gronow 2007.
5   Pohljobkin 1997: 49–50.

In 1947 the rationing system ended. Life in the cities was better than in the villages, where people continued to starve, so that there was an outflow of the rural population to the cities, leading to further urbanization. During the 1950s and 1960s, when the collective farms were 'amalgamated' to form larger units, fairly intensive voluntary, but to a large extent also, forced, migration took place from neglected and declining rural areas to the vicinity of agricultural centres. This movement changed the geography of ethnic composition significantly.[6]

During and after the war people had become used to public catering and, by the beginning of the 1960s, eating in public-catering establishments had overtaken the consumption of home-made food. Some catering establishments specialized in, for example, *bliny* (*Blinnaya*) or *pelmeni* (*Pelmennaya*) or tea (*Chaynaya*), and these became popular in factories, plants, and institutions, where food could be bought cheaply. There were 308,000 canteens, restaurants, cafés, and other public-catering establishments. The number of dishes made by them increased from 8 billion in 1940 to 51.8 billion in 1981.[7] At the same time, the art of cookery declined, because after the war many traditional national products and dishes were excluded from public catering, such as home-made fermented and salted vegetables, dried mushrooms, wild berries and jams. Many kinds of river fish disappeared, thus diminishing the range of fish dishes. Sea fish became available but for a long time did not find much response among the population. Young people lacking Russian culinary traditions were more used to eating meat dishes. One result of atheistic propaganda was to forbid the following of religious prescriptions, including fasting. During the 1950s and 1960s meat dishes prevailed over other dishes, and became a typical feature of the Soviet diet. Russian traditional cuisine had not included main meat dishes, so cutlets, tongue, escalopes, beefsteaks, schnitzels (based on fillet of pork) and other West European dishes of minced meat became 'Russian' for the first time from the 1960s to the 1980s. As a result, the present generation does not connect the concept of 'Russian cuisine' with fish and mushroom dishes.

Another new trend of Soviet culinary practice was pickling vegetables and fruits on a large scale. Home preserving or marinating was developed under the influence of the canning industries in Bulgaria, Hungary and the former Yugoslavia, which supplied canned food to the USSR. Vinegar, taste-improvers, pepper and the process of sterilization were used as the main means of preserving food instead of the traditional salting and fermentation in open utensils. The taste and also the composition of vegetable dishes (*paprikush*, tomatoes, capsicum-pepper, green peas, haricot and kidney beans and so on) that were used as garnishes differed considerably from the taste and composition of traditional Russian dishes, so that Soviet cuisine gradually gained a new character and taste. Between the 1950s and 1970s the consumption of eggs, poultry (especially broilers, chickens, turkeys, ducks), cooked meats and sausages increased sharply as convenience foods for main dishes. As to sausages, Soviet people knew only the so-called 'doctor's'

---

6   Voronina 1998: 246.
7   *About Products and the Culture of Consumption* 1984: 18.

boiled sausage by contrast with the Soviet élite who had special department stores with a variety of boiled and smoked sausages and other scarce foodstuffs.[8]

Nevertheless, Soviet cuisine was not formed at once. During a relatively short space of time – 70 years – it went through at least five stages, reflecting the socioeconomic history of the country. 'Soviet cuisine' is a cultural term that determines merely the culinary peculiarities established from the middle of 1930s until the beginning of 1990s. It had a special terminology of culinary art and a range of dishes that distinguished it from Russian traditional cuisine and from European or Asian cuisines.[9] While culinary borrowing is not new to Russian cuisine, which historically has included Caucasian, German and French dishes, the processes of adoption and adaptation were slower in earlier times. Even in the twentieth century, prior to the most recent economic crisis, Russian food products managed to retain their natural qualities.[10] Changes in the political, economic and cultural life of Russia during the 1980s considerably affected the traditional system of nutrition. The economic situation worsened and living standards declined. There was a so-called, 'mechanism of braking' crisis, when production indices fell to zero. From 1985, when Mikhail Gorbachev was elected as the head of the government, a new policy known as *perestrojka* was adopted. Gorbachev initiated a 'Food Programme' that needed more capital investment in the agrarian-industrial complex. At first, it was supported by many people, but the radical economic reform based on old socialist ideas did not occur and the situation reached crisis point. There were irregularities in the food-products distribution system and speculation in prices of the main products so that bread, sugar, salt and grains all went up in price. In 1990, the introduction of private property and the privatization of flats and houses gave rise to a new economic situation and, after an anti-state putsch in August 1991, the Commonwealth of Independent States formulated a new programme of reforms for the stabilization of the Russian economy.

New Russia – New Food – New Problems: 1991–2000

The disintegration of the former USSR in 1991 brought about a prolonged and severe economic crisis. Changes in the political life of Russia considerably affected the traditional food patterns. The food industry had difficulty in satisfying consumer demand. Its dislocation meant that many food products had to be imported from abroad. The main foodstuffs – bread, potatoes, and sugar – were available and people began to eat them in greater amounts, since products such as fruits and vegetables were in short supply. From the beginning of the democratic reforms in the 1990s, most people had to eat food high in kilocalories because of low living standards.

---

8   Pohljobkin 1997: 50–51.
9   Pohljobkin 1997: 46.
10   Voronina 1998: 249.

In an attempt to deal with the food shortages, radical liberal reforms were introduced. A 'shock therapy' method, introducing a free-market economy through the liberalization of market prices between January 1992 and December 1993, was intended to jolt the home economy into action. The immediate result was, unfortunately, otherwise. What actually happened was that the already almost unprecedentedly difficult economic situation was intensified, leading to a sharp drop in industrial production and a further lowering of living standards for most of the population.

The forced migration of ethnic Russians from the former Soviet Republics, especially from Middle Asia, was a result of the disintegration of the former USSR. They brought many dishes from the Republics to the central part of the country. Arising from the catastrophic economic situation of the Russian Federation in the early 1990s due to its political upheavals, foreign foodstuffs began to flood into the Russian market. In fact, as far as Russia was concerned at that time, it is probably more correct to speak of a 'food invasion' rather than 'food migration'. The question needs to be asked how this was received, and what influences exerted – positive, negative or creative – by this food torrent.

Thus at the end of the twentieth century, the definition of ethnic traditional food had changed in the context of Russia, and the movement of people, comprising ethnic migration, as well as food migration, played no small role in this change.[11] By contrast to the adverse food-supply situation for the waves of ethnic migrants described above, and for the Russian population in general in recent times, a specialized food sector has arisen in order to service the growing private tourist industry, which is attracting not only foreign, but also Russian tourists. The latter belong to a special social group called the 'new Russians' who, as bankers, international traders and so on, have achieved wealth and prominent social positions and, thanks to their foreign contacts, many new dishes have entered Russian cuisine. Accompanying the development of the private tourist industry has been the opening of prestigious and expensive gastronomic establishments offering the finest international cuisine, including coffee, pastries and desserts in the best Viennese tradition, as well as a wide variety of hitherto unfamiliar fresh and salt-water fish imported from many parts of the world. However, a result of this trend towards the provision of an international cuisine has been the disappearance of traditional Russian dishes from the menus of some catering establishments to the surprise of many tourists.

The economic difficulties at the beginning of 1990s generated steep inflation, and the way of life and the living wage changed very much. When the Russian economy stabilized, it became easier to have food in plenty and to choose anything from the shops. However, the fragmentation of society led to a great difference between the poor and rich people. The recovery of the economy was accompanied by a new stratum of citizens engaged in business, banking, and trade. Manual labour diminished considerably and this led to an increasing number of overweight

---

11   Voronina 2001: 70–72.

people. Obesity is slowly becoming a contemporary symbol of society, leading in turn to the popularity of Western ideals of slimness, good looks and low-calorie diets. This situation is more common for the well-to-do people in big cities, who pay more attention to their nutrition. It has become a mark of prestige to be known to be following a fashionable diet as part of the search for good health and beauty. At the beginning of the twenty-first century new diets spread quickly among different strata of the population. There are many articles about diets in newspapers, magazines and programmes on television, where famous people demonstrate their slim figure, low weight, beautiful face and body. In the mind of the Soviet people the word 'diet' was associated only with special medical treatments given free somewhere in a sanatorium, a workers' boarding house or convalescent home at the expense of trade-unions. Medical diets, elaborated by Ministry of Public Health, were also widely used for many public-catering establishments.[12]

A special study made by Saint Louis University (USA) in 2000 showed that the inhabitants of Eastern Europe and the former USSR began to get fat when the economies of their states went over to free market rules. By then the number of the people in Western Europe who suffered from obesity was smaller than in the countries of Soviet block. Obesity affected Norway only slightly, with only 6 per cent of Norwegians abnormally overweight. The problem affected Hungary (19 per cent) to a greater degree, and Russia and Lithuania (both 18 per cent) were also discussed by The Washington Profile Information Agency. The authors of the investigation determined which factors promoted obesity among people, taking into account the type of economy, quality of the government, the most popular diets, level of urbanization, availability of cars, quality of the roads, and so on. They concluded that the following reasons were significant: psychological stress connected with the most recent reforms, change of working conditions (for example, sudden increase in the number of workers engaged in offices and connected with computers), the expansion of fast food systems (for example, McDonald's), and the appearance of numerous cafés selling cheap pre-packed products such as sandwiches, which are very convenient for a quick snack.[13]

There has been a tendency to use partly-prepared foods, such as *Rollton* – dried broth or purée – to prepare a meal very quickly, but they include modified starch, boosters of taste and smell (sodium glutamate, artificial flavours identical to natural ones, vegetable oils). The firm going by the name 'Cuisine without Boundaries' also distributes partly-prepared food as a 'business-menu', 'business-lunch' or 'big lunch' but it is difficult to believe that the 'meat does not contain any preservatives and is ready for use' for, as their slogan says, 'these and other fast foods are popularized by mass media, especially by TV'.[14] A twenty-four hour delivery system to the client's home or office address became common through the internet or by distribution of leaflets showing different menus. Italian cuisine is

---

12 *Collected Recipes on Dietetic Nutrition for Public Catering Establishments*, 1989.

13 NEWSru.com, News of Russia, Wednesday, 7 June, 2007.

14 Source: http://www.kbgfood.ru/ [accessed: 14 February 2009].

the most popular, followed by Japanese in second place. It is interesting that, after a long interval, Russian traditional food has regained popularity and it is easy to order *borshch*, *bliny*, porridges, salads and other dishes through the internet.

With regard to weight-loss, nutritionists recommend those diets that help to reduce extra fat and normalize bodyweight by various special methods and food rations. There are 'poly-component' diets (meat, vegetables) and 'mono-component' diets when only one group of products can be eaten (vegetables or fruits or meat or even products rich in carbohydrates). The 'selective', 'naturopathic', 'contrast' and other diets are also popular. A new 'artificial' diet that is based on eating special bioactive additions (BAD) and artificial food products gives very quick results but may damage health. There are other nutritionists who have a contrasting opinion about these diets. For instance, V. Mihajlov, the author of a book about the danger of separate diets, has popularized the theory of rational and balanced nutrition. He thinks that it is necessary to analyse the various diets and to approach the question rationally rather than following blindly the recommendations of fashion-conscious dieticians. He also suggests recipes for dishes that agree with contemporary scientific opinions on nutrition.[15]

Losing weight requires using stored fat as an energy source. A traditional way is to create an energy deficit – that is to say, 'to go on a diet'. In Russia, the Kremlin diet became very popular at the end of the twentieth century and the beginning of the twenty-first century. It gained its name thanks to the Kremlin politicians or members of the Russian government who started to follow this regime. It is also called the 'Rublev diet' because of the Rublev highway to a place near Moscow where politicians and other famous people live, and also to other places near Moscow and Saint Petersburg.[16] This diet is followed by wealthy people who can make enough time to follow the principles of the diet and attend massage and fitness clubs. On the whole the Kremlin diet is thus a diet for the well-to-do stratum of the population or the élite of society, including the heads of firms, businessmen, actors, and writers, whom ordinary people call the 'New Russians'. They enjoy a night life in clubs, casinos, parties and meetings because of their professional interests. It is interesting to note that their diet includes many foreign dishes and products, for instance, alcohol, wines and sea food. Among the many available diets, the Kremlin diet is the most paradoxical. Indeed, to become slim it is necessary to eat meat and – most impressive of all – it allows the drinking of strong alcohol. That is why a keen feeling of hunger that hinders work does not occur on this diet.

The main point of the Kremlin diet is that it permits eating meat, fish, cheese and vegetables, but it avoids products high in carbohydrates. This diet is not suitable for those with a sweet tooth; indeed there is only one piece of sugar in the daily ration. The Kremlin diet restricts all sweet, floury dishes, dishes made with potatoes, grain and also juices, fruits and vegetables high in carbohydrates.

15　Mihajlov 2003: 7–8.
16　Pishalev 2006: 3–5.

All the products in the Kremlin diet are allocated marks – 1 mark is equal to 1 g of carbohydrate. It is necessary to keep to 40 marks a day to lose weight and to maintain a low weight. Weight is put on when eating totals over 60 marks a day. A great deal can be eaten but it is better to follow suggested quantities. It is not good to eat 3–4 kg of meat a day, then 1 kg of cheese, while drinking 1 litre of table wine, although the marks in this system allow it! The main principle of the Kremlin diet is to accumulate only a few marks but to eat enough. There is a special table of the 'nutritional' value of some products in marks. For example, meat has a mark of zero, fish is also zero, so they can be eaten in plenty, but fruits and vegetables have more marks: apricot – eight, orange – seven, water melon – eight, banana – 21, peach – nine, plums – nine, apple – nine.

However, there is an opinion that the Kremlin diet is not healthy, and the only alternative is a balanced diet, eating everything 4–5 times a day in small amounts which should encourage the loss of 500–800 g weight a week. The metabolism must remain as a normal and natural process.[17]

The wide distribution of new diets in contemporary Russia requires consideration of the circumstances that have caused the numbers of overweight and obese people to increase every year. It seems strange for a country where half of the population do not eat enough: there are 62 million poor people and 290,000 rich, amongst whom there are 90,000 millionaires. The Russian Centre for the Study of Living Standards carried out research in 2005 that showed there are seven strata in the population with different levels of income. In 2006 an official Russian 'nutrition basket', that is a set of the main food products needed to survive at a subsistence minimum, was adopted by the State Council. It included monthly quantities of 10 kg of bread and floury items, 8 kg potatoes and but only 1 kg of vegetables. Table 3.1 shows the changes in the subsistence minimum in Russian roubles for the first and second quarters of 2007.[18]

Table 3.1   Subsistence Allowances in Russia, 2007

Allocation per head	1st Quarter 2007 (Roubles)	2nd Quarter 2007 (Roubles)
Average overall	5,609	5,772
Able-bodied person	6,360	6,533
Pensioner	3,830	3,967
Child	4,780	4,936

17   *Anti-Kremlin Diet* 2005: 59.
18   These data are given monthly in many newspapers; the most recent ones were taken from the *Moscow Komsomolets*.

Table 3.1 shows that the poverty-line income rose by 163 roubles. Taking into account that one United States dollar was equal to approximately 25 roubles, the subsistence wage for those who able to work was 261 dollars a month. This gives an idea of an ordinary salary in Russia.

In the 1980s, there were only eight per cent of people who suffered from obesity, but in 2003 the proportion had risen to more than 21 per cent. Therefore diets are in demand now, and taking into account that they are popularized in many verbal and written forms, one can say that a diet industry has been born. On the one hand, this can be explained by the fact that, following the democratic reforms, Russia became more open to the West European and American markets, leading to the availability of much imported food. On the other hand, new groups in the population like the 'New Russians' – businessmen and especially people engaged in banks (the number of the banks has considerably increased) – have a different working environment and lifestyle. For them, it was a big problem in the 1990s whether to have lunch or snack because the old system of public-catering establishments was in ruins and only a few new ones replaced them. Having more money than ordinary people, high and low ranking officials visited the restaurants and cafés that had also changed in accordance with new demands. It became fashionable to visit restaurants after work and to relax in the company of colleagues or friends. As the new élite became more engaged in international business and affairs, they began to visit other countries and to attend official functions. Coming into contact with different surroundings and people, the new Russians began to follow the widespread norms of behaviour and outward appearance.

The new Russians eat mainly foreign food products and some of their habits promote obesity. These include over-eating, smoking, drinking alcohol, a liking for coffee and refined products, eating too much salt, processed food products, and the consumption of animal proteins. In Russia, the problem of obesity is now nearly as serious as the problem of alcoholism. One of the reasons is the fact that the contemporary market is full of different 'fast foods' or 'junk foods' rich in fats and sugar, like doughnuts, sticks of confectionery, Big-Macs, hamburgers and other burgers of different types, chips, snacks, nuts and so on, that give the human body energy. Targeted attractive advertising is partly responsible for a growth in children's obesity.

New stereotypes of beauty that have been imported into Russia have become necessary for happiness, success and self-respect. A fashion to be slim very often goes with the problem of overcoming eating disorders and stresses.[19] The statistics demonstrate that every second woman between 18 and 29 years old tries to lose weight, even when there is no clinical need to do it. The consequence of such un-controlled slimming and the fear of weight-gain may be anorexia nervosa – an illness that results in a loss of appetite and depression. Related to anorexia is bulimia nervosa – binge eating followed by vomiting. Psychiatrists confirm that solitude, absence of love, stress, a feeling of restlessness, unrealizable dreams, a

---

19   Brigham 2003: 3–4.

fear of life or death, can be the basis of many problems and diseases, including obesity. To a certain degree food compensates for some of these factors. Other factors, such as a sedentary lifestyle connected with certain professions and with the use of computers, also play a significant role in weight gain.

Conclusion

The changes in Russian eating patterns in the twentieth century depended on many influences. First of all, political circumstances played an important role at every stage of the country's history. The changes in everyday life during the Soviet period led to a standard food ration that was reflected in fewer dishes of the traditional cuisine surviving and the introduction of new recipes. On the whole, although there are many influences on the everyday nutrition of contemporary Russians, the main factors include those changes in the world that have led to the globalization of food supply and the emancipation from culinary problems at home.

Democratic reforms in the 1990s brought many difficulties into everyday life. Different living standards adversely affected the food intake of the greater part of the population and led to an increased consumption of more bread, floury and other foods of poor quality. This has meant a wide dissonance between different strata of the population and this disparity is growing. Among the new stratum of citizens engaged in business, banks and trading and who enjoyed eating dishes of high quality, there has been an increasing number of overweight people. So, at the end of the twentieth century and the beginning of the twenty-first century, the problem of obesity in Russia was transformed from a few individual cases into an urgent and widespread social problem. Nutritionists of various opinions are now engaged in inventing new diets to help people lose weight. The extent of the condition has resulted in a situation where the struggle with extra weight has become a kind of business. On the whole, the tendency to be overweight among Russians reflects a real socioeconomic situation of contemporary society and proves that an abundance of foodstuffs does not always lead to a healthy life. Obesity, as a result of overeating, can lead to serious illnesses. An improvement in the nutritional quality of the diet in Russia must be combined with a universal educational process to provide information on how to eat and to expend energy. Considerable changes in the traditional system of nutrition in Russia led to the growth of serious multi-factorial illnesses of the endocrine and cardiac systems. So, it is necessary to advise about possible illnesses and to provide informed choices for both adults and children.

References

About Products and the Culture of Consumption [O productah i culture potreblenia], Moscow, 1984.

Anti-Kremlin Diet or Revolution in Nutrition [Antikremljovskaya dieta ili revoljutsija v pitanii], Moscow, 2005.

Brigham, S., *Stresses and Eating Disorders [Stressy i narushenija pitania]*, Saint Petersburg, 2003.

Collected Recipes on Dietetic Nutrition for Public Catering Establishments [Sbornik retseptur dieticheskogo pitania dlja predprijatij obshestvennogo pitanija], Moscow, 1989.

Ginzburg, M., *A Deficit of Energy [Deficit energii]*, Moscow, 2006, 124–8.

Gronow, J., 'First-Class Restaurants and Luxury Food Stores. The Emergence of the Soviet Culture of Consumption in the 1930s', in ICREFH IX, 143–53.

Mihajlov V., *Towards a Harm of Separate Diet [O vrede razdelnogo pitania]*, Moscow, 2003, 7–8.

Pishalev, V., *Rublev Diet of Our Elite [Rublevskaya dieta nashej eliti]*, Moscow, 2006, 3–5.

Pohljobkin, W.V., *National Kitchens of our People [Natsionalnye kuhni nashih narodov. Povarennaya kniga]*, Moscow, 1997.

The Book About Tasty and Healthy Food [Kniga o vkusnoj I zdorovoj pishche], Moscow, 1939.

Voronina, T.A., 'The Impact of Political Change on Ethnic Food in Twentieth-Century Russia', in Lysaght, P. (ed.), *Food and the Traveller: Migration, Immigration, Tourism and Ethnic Food*, Nicosia, Cyprus, 1998, 246–52.

Voronina, T.A., 'Cereals in Everyday, Festive and Ritual Life of Russians' (*Zernovye v povsednevnoj, prazdnichnoj i obrjadovoj zhizni russkih*)', in *Bread in Popular Culture. Studies in Ethnography [Hleb v narodnoj culture. Etnograficheskie ocherkee]*, Moscow, 2004, 101–38.

Voronina, T.A, 'Traditions in the Russian Food Patterns at the Turn of the XX–XXIst Centuries' [*Traditsii v pishe Russkih na rubezhe XX–XXI vekov*], *Traditional Food as an Expression of Ethnic Identity [Traditsionnaya pisha kak vyrazhenie etnicheskogo samosoznania]*, Moscow, 2001, 41–72.

Chapter 4

Slovene Food Consumption in the Twentieth Century – From Self-Sufficiency to Mass Consumerism

Maja Godina Golija

At the beginning of the twentieth century, Slovenia was predominantly an agrarian country, with most of its population living in the countryside and tilling the land. There were few large towns. Although gradual industrialization after the First World War changed this situation somewhat, 66 per cent of Slovenes still worked in agriculture in 1921. In 1931, this percentage had fallen to 59.5 per cent and gradually then to slightly over 50 per cent in 1940.[1] Only larger towns such as Ljubljana and Maribor had big industrial plants that drew emigrants from nearby villages and accelerated urban growth. There were also some important mining and iron ore centres, for instance Trbovlje and Laško.[2]

Generally self-sufficient, Slovene farmers worked the land for their own household needs and to a lesser extent to market their crops.[3] Mostly growing crops indigenous to where they lived; they provided a variety of produce typical for the geographically diverse regions that make up Slovenia. It is therefore possible to say that this was a time when the food culture in Slovenia was still very much geographically differentiated and staple foodstuffs did not yet come from traded goods sold in shops.

According to an ethnological classification, there are four major types of food culture in Slovenia.[4] The Pannonian type in the east is based on crops like wheat and buckwheat. Meals made from wheat and buckwheat flour consisted of different types of bread, pasta, and pies often filled with cottage cheese. Dishes were flavoured with sour cream and cottage cheese. Abundant crops of pumpkins, not grown anywhere else in Slovenia, gave excellent pumpkin oil widely used in cooking.

---

1   Cvirn 2000: 339.
2   Slovenia was a province of the Austro-Hungarian Hapsburg Dual Monarchy until 1919. It was then part of the Kingdom of Serbs, Croats and Slovenians which became Yugoslavia in 1929. Yugoslavia was a republic from 1945 until 1991, when Slovenia became independent.
3   Ložar 1944: 98.
4   Godina Golija 2006: 51.

The northern, or Alpine type, typifies the hills, mountains, and forest areas. With the exception of corn and buckwheat, its harsh climate does not provide adequate conditions for agriculture, but is suitable for animal husbandry and Alpine dairy-farming. The food culture of this region was thus based mainly on dairy products such as milk, sour milk, curds, and cheese, and corn and buckwheat porridge. Venison, which was rarer in other parts of Slovenia, could also be found on the tables of local households. Meat from game animals was also cured and made into sausages and other meat products.

The central type characterized middle Slovenia. Its food consisted of staples that grew best in this climate. People planted tuberous vegetables such as potatoes and turnips, in addition to buckwheat and millet. Buckwheat and millet porridge, boiled in water or milk, was prepared frequently, as were cabbage and turnip. This was the first Slovene region whose population started to include the potato in their daily meals; potatoes quickly became very popular and were prepared in a number of ways.[5]

With its warm Mediterranean climate and karstic, limestone, soil, western Slovenia gave birth to a Mediterranean type of food culture.[6] Rather than supporting cereal cultivation, the barren soil is suitable for olive trees, certain kinds of vegetables and fruits, vines, and for raising sheep. Among the most widely served foods, included in most meals, were polenta (a bread substitute), thick vegetable soups called minestrone, vegetable and meat sauces, fish, and the widely used olive oil.

Changes in food culture were introduced gradually and were connected with the growing mobility of the rural population. Whether working in towns on a regular basis or only occasionally, this segment of the population adopted some urban food customs and transplanted them to their original rural environment. Due to better road and rail connections, a food trade was made possible to smaller and more remote places. It was also possible to seek employment outside the home village or in another town. Dependency on home-grown food lessened, which in turn denoted that food became more and more uniform and started to lose its regional characteristics and peculiarities.[7] More perceptible changes, however, started at the end of the 1950s and especially in the 1960s when the rising standard of living in Yugoslavia resulted in an increase of its population's purchasing power.

The situation in Slovene towns was similar, with food supply still largely dependent upon the produce from nearby farms. Urban Ljubljana, Maribor, Celje, and Trieste all had thriving rural communities in the vicinity that provided a steady supply of produce carried to the towns by farmers. Until the middle of the twentieth century, the majority of the Slovene urban population obtained their food mostly from nearby agrarian areas. Towns also had shops selling imported

---

5   Godina Golija, 2006, 51.

6   Karsk, near Trieste, is on a bare limestone plateau which has given its name to the geology of the whole region.

7   Godina Golija 2006: 52.

foods and spices from distant lands called colonial-goods shops; they were more expensive and therefore hardly accessible to most Slovenes. Market days presented an opportunity for farmers and retailers from more distant places to sell their produce to town residents. They arrived by train, wagon, and even on foot. *Šavrinke*, the Istrian women who sold produce in Trieste, sometimes lived so far away that it took them several days to walk there. They also bought produce such as eggs, for instance, from farmers living along the way and sold them in open-air markets or door-to-door.[8]

As in the First World War, food shortage and irregular food supply characterized the period just before, during, and after the Second World War. Although ration cards were introduced in the interwar period, food supply was still disorganized. In his memoirs of the First World War, Dr Avgust Reisman, a lawyer from Maribor, says that the front was not only on the battlefield, but also at home, with long lines of people queuing in front of food offices and later in groceries, butchers' shops, and bakeries.[9] According to interviews with people with first-hand experience of the First World War, ration cards did not help much in themselves; in order to obtain the desired victuals it was necessary first to win the goodwill of bakers, grocery owners, and butchers. Protests of enraged town residents were therefore not infrequent.[10] Especially stressful was the shortage of flour. It was strictly forbidden to bake buns and other fancy bread during the war; bakers were only allowed to bake bread in loaves, but these could not be made of fine wheat flour. Confectioners were not allowed to use more than 30 per cent of wheat flour for their pastries. Bread products which thus included ingredients such as ground pine cones, sawdust, or chestnuts, tasted bad, had a strange smell, and did not keep long.[11]

During the war, townspeople tried to improve their diet by obtaining victuals from their relatives or acquaintances living in the country. In spite of severe restrictions, it was sometimes possible to conceal foods and sell them in the black market later. Written sources mention crowds of people setting out for the country on Sundays, among them smartly dressed ladies who were not in the least ashamed to carry heavy knapsacks crammed with food back into town.[12]

Soon after the First World War the food supply became more regular. In November 1918, Slovene towns set up municipal food councils responsible for the regular supply of food. The National Government in Ljubljana determined prices for basic staples such as wheat, rye, oats, millet, buckwheat, potatoes, and kidney beans. It unfroze the ban on pig slaughtering and the fruit and vegetable

---

8   Ledinek 2000: 25–7.
9   Reisman 1939: 54.
10 Godina Golija 1996: 34.
11 Godina Golija 1996: 35.
12 Reisman 1939: 61.

trade. Several months later, newspaper articles stated that the food supply was once again adequately restored and that food prices were normal.[13]

With the change in economic circumstances as a result of the newly-established Kingdom of Yugoslavia and the ensuing changes in crops and produce market, food prices constantly fluctuated throughout the 1920s and the 1930s. In view of these changes, the government issued an official expenditure record of necessities of life for two types of three-member family: for the middle-class urban family and for the working-class family with the monthly expenditure of 1,077 and 549 dinars respectively.[14] The expenditure of a working-class family included 10 kilograms of bread flour, 3 kg of Indian meal (polenta), 1.3 kg of beef, 9 litres of milk, 6 eggs, and 1 kg of sugar.[15] The urban family's expenditure included the following indispensable household items: 12 kg of white flour, 1 kg of Indian meal, 1.80 kg of beef, 0.90 kg of veal, 0.80 kg of pork, 15 litres of milk, 14 eggs, and 1.80 kg of sugar.[16]

A comparison of these items shows that both family types used a substantial amount of flour. This is understandable since rural and urban families alike still baked their own bread at home. Expenditure calculations presumed that working-class families did not have to buy vegetables and pork in shops, but grew their own vegetables and raised pigs. Nine litres of milk per month, on the other hand, was hardly enough for a working-class family.[17] Informants still remember that milk was scarce and often mixed with water or, usually in less affluent families, substituted in the morning by soups such as one made of browned flour, water, and spices.[18] Consumption of sugar was relatively high.

A comparison of purchases that had been entered into expenditure notebooks in the 1950s indicates that the structure of provisions did not change essentially after the Second World War as Table 4.1 shows. Flour was still the most important item on the list because bread was still mostly baked at home. Almost as important were lard and coffee substitutes. Sugar consumption was still high, which indicates that sweets were mostly made at home and that housewives also bottled fruit themselves.[19] This list already includes items that prior to that time had not been served regularly, but were incorporated in Slovene daily menus in the 1950s and the 1960s, such as tinned food, especially canned sardines and tinned pâté.[20] Pasta, vegetable oil, cocoa, different kinds of salami, and bottled mineral water were also purchased much more frequently.

---

13   Godina Golija 1996: 35.
14   Mlakar Adamič 2004: 28.
15   Mlakar Adamič 2004.
16   Mlakar Adamič 2004.
17   Mlakar Adamič 2004. Neither veal nor pork was included for the working-class family.
18   Godina Golija 1996: 53.
19   Mlakar Adamič 2004: 29.
20   Mlakar Adamič 2004.

Table 4.1   Monthly Food Consumption per Family (4 persons)

Survey	Flour, groats (kg)	Meat (kg)	Milk (kg)	Eggs (number)	Sugar (kg)
1936	13.0	2.4	12.0	10.0	1.4
1954	31.1	6.8	33.7	30.0	5.3
1961	8.9	6.5	35.8	30.0	6.9
1973	3.5	9.2	31.5	33.2	4.9

Source: Mlakar Adamič 2004.

During most of the twentieth century, the food supply and consumption of different foods was in proportion to the low income and purchasing power of the Slovene population. This is indicated both by the expenditure data for an average Slovene household and by prices of individual articles. In the period prior to the Second World War and in the years just afterwards, prices were so high that only a few could afford to buy food sold in shops. In 1935, at an average hourly wage rate, a Slovene worker had to work 23.3 hours to earn the price of a kilogram of coffee; 4.66 hours for 1 kg of pork; 6.78 hours for 1 litre of cooking oil; and 5 hours for 1 kg of sugar. By comparison, only 0.33 hour was needed to earn enough to purchase 1 kg of potatoes, and 0.83 for 1 kg of flour.[21]

Introduction of Changes in the Food Supply

United Nations Relief and Rehabilitation Administration (UNRRA) packages, first distributed in November 1945, brought about certain changes in the traditional food culture of Slovenia. The first packages contained biscuits, a can of pressed ham, a can of pressed bacon, dried codfish, cheese, cigarettes in packages of five, several packages of coffee, packages of tea, canned milk, roasted peanuts, peanut butter, milk powder, marmalade or jam, chewing gum, packages of chocolate or cocoa, and sugar cubes. Cod-liver oil was dispensed among the population as well.[22]

The distribution of UNRRA food packages was organized by the Red Cross and by social welfare offices. Most were distributed to schools, kindergartens, and hospitals; each family received several packages per year. Packages differed in size and contents. Large ones were issued to labourers engaged in heavy work, such as miners, middle-sized packages to skilled workers, for instance carpenters, mechanics, bricklayers, and so on, and small ones to unskilled labour. Beside the items mentioned before, UNRRA packages contained foods hitherto

---

21   Godina 1992: 75.
22   Mlakar Adamič 2004: 38.

unknown to Slovenes: sweet and salted crackers, corn in brine, fruit bars, and fish puddings. Foods such as canned bacon, for instance, which tasted similar to what Slovenes were already used to, were well liked; unfamiliar ones caused disappointment and rejection, and were even fed to pigs. Never seen before and therefore especially baffling, were powdered milk and powdered eggs. People had no idea how to prepare them correctly, especially since instructions were only in English. Powdered eggs were quickly nicknamed *Trumanova jajca*, which literally translates as Truman's eggs (but really means 'Truman's balls') named after the American President, Harry S. Truman.[23] These eggs were not consumed. Similarly unfamiliar with the correct preparation procedure for powdered milk and powdered coffee, people mixed them with some water, added sugar, and ate the thick concoction as a pudding.

Since severe food shortages persisted until the end of the 1950s, food from UNRRA packages was also sold in shops. With the economic growth of the 1960s and the 1970s came a stable food supply. People could also buy imported foods, for instance, such as citrus fruits, bananas, chocolate, tea, and coffee; yet the selection was still somewhat limited. As elsewhere in socialist Yugoslavia, there were occasional shortages of these imported products at the end of the 1970s and at the beginning of the 1980s; this was due to the depletion of foreign currency reserves which limited imports. Especially difficult to obtain were coffee and chocolate, which is why some people bought them in neighbouring Austria and Italy, or from profiteers. After several years, this situation normalized. When Slovenia seceded from Yugoslavia in 1991, foreign chain stores like the Austrian Spar, the German Hofer, or the French Leclerc were quick to open outlets all over Slovenia. This gave rise to a new phenomenon, the so-called hypermarket, which offered an infinite variety of foods by different manufacturers and fruit and vegetables from various suppliers and from many parts of the world; they also sold household articles. In the face of such fierce competition, some smaller Slovene commercial enterprises were bankrupted and local grocery shops operating in residential or in rural areas had to close down. Some large Slovene enterprises, such as Mercator and Tuš, for instance, adapted to new circumstances and started to open their own hypermarkets.

Shopping for Food

Since Slovenia has extensive agricultural areas that in the past yielded most of the produce needed for its population, only rare products had been purchased in open-air markets or shops. Trade and the export of agricultural produce to neighbouring lands was also an important economic factor. Farmers, who produced mostly wine, olive oil, vegetables, and fruit, sold them in nearby towns or exported them by

---

23   Apart from its usual meaning, the Slovene term *jajca* is also used as a derogatory synonym for testicles.

train as far as Vienna and Trieste. Amongst the most renowned were cherries and peaches from Goriška Brda, apples from Štajersko, and early spring vegetables from Primorsko.[24]

Some victuals were bought in grocery shops that were in operation until the 1960s. They were especially numerous in towns, but could also be found in larger villages. Residents of towns could also buy food in the so-called 'colonial' shops that sold a large selection of imported teas, coffees, spices, and chocolate, and also at greengrocers, bakeries, dairies, and in butchers' shops.

Prior to the Second World War, urban and mining centres also had purchasing and consumption cooperative societies.[25] Since they were exempt from paying sales tax, their prices were lower, thus competing with grocers. Their customers were low-income people such as labourers and junior clerks, while senior civil servants and even mayors' families bought their food at civil servants' purchasing cooperatives.[26] In consumption cooperatives, goods were usually purchased on credit; customers had to settle their accounts on paydays. At the end of the year shopkeepers added up the sum of each buyer's yearly purchases and rewarded their clientele accordingly; this was a common procedure in other shops as well, both in rural areas and in towns.

At the beginning of the twentieth century, most housewives did their grocery shopping once a week or once a fortnight, most often on paydays. Reckless, and therefore usually more expensive and often unnecessary, purchases could be avoided by careful planning. Costly delicacies or ingredients for a festive meal were bought only occasionally. A typical weekly purchase of food, which consisted of only the most basic items necessary to prepare meals during the week, included flour, salt, Indian meal, grits, coffee, sugar, oil, vinegar, lard and certain spices; items which were indispensable in the everyday preparation of meals. These were affordable despite the generally low purchasing power of the majority of the Slovene population. Food prices depended on their quality. In the mid-1930s, 1 kilo of flour cost anywhere between 2.5 and 3.5 dinars; 1 kg of salt from 2.8 to 3 dinars; and 1 kg of Indian meal or grits from 2 to 4 dinars. Coffee, rarely drunk at home, was sold in very small packages usually containing between 30 to 50 g; it was sparingly added to coffee substitutes to improve their taste. Coffee substitutes were made of roast barley, chicory, and coffee substitutes manufactured by the Kneipp or Franck companies. The cheapest brand of sugar was sold in small cones and called 'štokdeker'. Pricier was granulated sugar that cost from 13.3 to 15 dinars per kilo; at 16.5 dinars per kilo, sugar cubes were the costliest.[27] Although prohibited, saccharin was used for sweetening certain dishes and beverages. Less affluent households obtained it from hawkers selling it door-to-door.

---

24   Sketelj 1998: 71.
25   Adamič 2004: 25.
26   Godina Golija 1996: 38.
27   Godina Golija 1996: 39.

Prior to the Second World War, bread was mostly baked at home and not bought from bakers. Urban housewives customarily kneaded it at home, and then took it to a nearby bakery for baking. Costlier kinds of bread such as buns, rolls, and milk-roll croissants were only eaten in wealthier families, even then very rarely and only on very special occasions by other social classes.

Urban residents bought meat from butcher's shops and meat stalls in open-air markets. There was a wide selection of meats that cost more in 'fancier' butcher's shops, but were cheaper and of second-class quality when obtained from wholesale meat merchants in the suburbs. Live poultry as well as horse meat was available in town markets. Yet most meat did not come from shops. Pigs, which were habitually slaughtered in the days before Christmas, were reared not only on farms, but also in towns. Poorer families raised rabbits and goats whose meat was often served on Sundays.

Most urban households bought milk and dairy products such as cheese, cream, butter, eggs, and also oil, from farmers living in the vicinity. Every morning they delivered milk to individual households as well as numerous town merchants; these could also buy milk from big landowners specializing in milk production. In the countryside, farmers usually obtained milk from their own livestock. Seldom consumed fresh, it was ordinarily turned into dairy products, especially butter, which represented an important source of income for rural housewives.[28]

In the first half of the twentieth century, Slovenes generally did not buy fruit or vegetables, but grew them in their own gardens and orchards. Rural houses usually had a vegetable garden and an orchard. Many houses in town, including rented apartments, had small gardens or pieces of ground that could be rented together with the apartment; some people even rented whole fields for growing vegetables and potatoes. Only those living in the town centres proper (which were often of medieval origin) had no gardens of their own and had to resort to buying fruit and vegetables in markets or from costermongers. Some housewives from affluent burgher families frequented the market every single day. The most important market days before the Second World War were Friday and Saturday, with the largest selection of goods. While garden produce was laid out on tables, fruit, including citrus fruit, dairy products, and eggs could also be bought from carts positioned in the market area. Cages with live poultry, especially chickens and hens, stood on the floor, next to crates with turnip, potatoes, and legumes like broad beans and kidney beans. Markets provided a wide variety of goods in all Slovene towns; the sole exception was the winter of 1940 when a bad harvest and severe cold caused a shortage of vegetables and potatoes.[29]

It was not yet common for vegetables and fruit to be weighed on scales but they were sold by the piece or in small heaps; potatoes, walnuts, flour, and nuts and berries from the forests were also sold by measuring cups. This provoked a number of complaints in daily newspapers. Unable to supervise this rather imprecise

---

28   Godina Golija 1996: 41.
29   Židov 1994: 62–3.

weighing process closely, customers often felt cheated and disappointed. It was not until the Second World War, when the occupying authorities introduced food rationing, that goods started to be weighed. Postwar years brought new purchasing habits.[30] After ration cards were abolished in 1953, the newly set-up unions and workers' associations were in charge of organizing regular food supplies and obtaining provisions for their members. With the gradual decontrol of trading, people were finally able to buy food according to their earnings.

In Slovenia, the first supermarket shops selling food and other household goods on a self-service basis opened in the 1960s. The shelves of large Slovene commercial enterprises, such as Koloniale and Mercator, predominantly offered domestic food products that were selected by the buyers themselves and paid for at the cash register; this was certainly a novelty for Slovene consumers. During shortages of certain products, some families bought their groceries in nearby Italian and Austrian towns. After Slovene independence in 1991, the food market stabilized and brought in its wake large hypermarkets vying for customers in a number of ways. Just like food merchants and shop owners of a century ago, they attempted to lure their regular customers with discounts, sales, and small gifts. In present-day Slovenia, the grocery shop network is well-developed; on average, every third village has a grocery. These usually sell food products or assorted goods, while larger communities also have a number of other, more specialized shops.[31]

Preparation and Consumption of Canned Food

Housewives prepared some foods themselves. Prior to the introduction of freezers in the mid-1970s, people preserved meat in a number of ways. Bacon, whether obtained from pigs reared at home or purchased at the butcher's, was fried in lard to make crackling and minced lard. Once widely employed in cooking, lard started to be substituted by oil in the 1960s. Meat was also dried or smoked and, prepared in these ways, kept for a long time. Sometimes pieces of meat or sausages were first baked, then covered with lard and kept in special receptacles. Meat products were also wrapped in paper and covered with grain.

Sauerkraut and turnips, the most important winter vegetables, were prepared in most households. Some families made their own apple or wine vinegar. In order to preserve their freshness longer, eggs were kept in lime water. These provisions were made mostly in May and in August when the quality of food was at its peak and its price low. Home-made fruit and vegetable preserves did not become popular until the decades after the Second World War. This was probably due to the high price of sugar, the best-known preserving agent at the time, which many families could not afford. Sometimes they could not afford special glass containers needed

---

30   Godina Golija 1996: 43.
31   Fridl 1998: 240.

for preserves. Fruit preserves were made from a variety of fruits. Raspberries were made into syrups, plums, apricots, apples, and pears into jams and stews.[32] The most typical vegetable preserves, consumed in wintertime, were tomato paste and pickled gherkins and beetroot, but with the increasing popularity of freezers since in the 1970s many housewives started to freeze fresh produce instead of preserving it.

Industrially canned food, which was mostly imported and therefore very expensive, was very rarely purchased prior to the Second World War. Selections available in groceries consisted mostly of tinned fish such as sardines or tuna; butchers and meat-products companies also sold tinned meat. Shoppers were also offered pickled cucumbers and tomatoes, and plum jam; because of its firm consistency, the latter was usually thinned with water. In the years before the Second World War, tinned pre-cooked meat or other food products such as, for instance, beef stew, was purchased very rarely, and if so, then mostly tinned fish. These were extremely expensive and therefore affordable only for a select few families, and only on very special occasions for the rest. An informant, born into the family of a tannery-plant general manager, remembers that a tin of fish was so expensive that in her family it was only rarely served for dinner, when it was split between four members.[33] The development of the food-processing industry after the Second World War brought considerable changes in the food culture of Slovenes. Tinned food became affordable for the majority of the population. The consumption of tinned meat, especially meat or liver pâté, luncheon meat, and tinned fish, began to increase in the middle of the 1960s. Especially popular were pickled beets, green peppers, and olives, and mixed vegetable salads.

Income and Food Expenditure

According to statistical data for the period prior to the Second World War the average monthly income of an industrial labourer amounted to 800 dinars; most of which was spent on food, and the rest on lodgings, clothes, footwear, and other expenses. More detailed statistical data show that in 1953 the average yearly income in Slovenia was 18,503 dinars, of which 8,633 dinars – or 46.6 per cent – were spent on food; 2,898 dinars, or 16.7 per cent, on clothing and footwear; other expenses being much smaller.[34] In 1968, when the average Slovene yearly salary was 50,126 dinars, more than one-half covered food expenses (25,254 dinars or 50.4 per cent). This percentage rose to 50.5 per cent in 1972[35] when the average yearly salary amounted to 107,113 dinars, 54,108 of which were spent on food. This gradual, yet constant increase, shown in Table 4.2, indicates that

---

32 Godina Golija 1996: 46.

33 Godina Golija 1996: 48.

34 *Statistični godišnjak FNRJ* 1954: 310.

35 *Statistični godišnjak* 1973: 316.

more food was purchased in shops and less grown at home. In 2002, after eleven years of Slovene independence, the average yearly income per family member was 1,118,761 Slovene tolars, 211,900 (17.4 per cent) of which went toward food expenses.[36] This indicates that food expenses decreased considerably, meaning that, in comparison to other items, food became cheaper. Most Slovenes now purchase food in shops instead of growing their own; if they do, it is mostly because they wish to grow organic or at least more wholesome produce that tastes better and is healthier; economic factors are less important.

Table 4.2   Percentage of Average Yearly Income Spent on Food

Date	Annual Income (dinars)	Food Expenditure (dinars)	Percentage (%)
1953	18,503	8,633	46.6
1972	107,113	54,108	50.5
2002	1,118,761	54,108	17.4

Source: *Statistični godišnjak FNRJ* 1954, Godina Golija 1996, 46–8.

Note: In 2002, the Slovenian currency changed to the tolar.

Conclusion

An analysis of food supply in twentieth-century Slovenia shows that until the last decades the majority of the Slovene population depended mostly on produce they could grow themselves. Rather than purchasing them in open-air markets or shops, vegetables and fruit usually came from domestic gardens; after a comparison between income, food prices, and the general standard of living in Slovenia, this is perfectly understandable. The standard of living, which was much lower than in more developed European countries, hardly allowed for any extensive purchases of shop-bought food and its use in the kitchen. Some of the key factors for this were that the population was mostly rural, women were usually not employed outside their homes, and families were much larger.

Research indicates that at the beginning of the twentieth century, the most frequently consumed foods in Slovenia were bread and other farinaceous products; these were followed by milk and other dairy products, then vegetables, meat, and fruit.[37] This structure remained more or less the same after the Second World War. Statistical data show that Slovenes still consumed mostly bread and other farinaceous foods but the consumption of potatoes, milk and eggs was also large. Meat, vegetables, and fruit were consumed in smaller quantities. Monthly

---

36   *Statistični letopis Republike Slovenije* 2004: 271.
37   Mlakar Adamič 2004: 28.

sugar consumption was still large (5.3 kg).[38] These data are further corroborated by expenditure data on provisions from the 1960s and the 1970s. The largest part of the monthly income was spent on bread and other farinaceous food, meat, and milk; less money was spent on vegetables, fruit, sugar, and sweets.[39] Statistical data from the last two decades, however, indicate changes in the structure of meals and food purchasing. In 2002, the average Slovene household spent 3.4 per cent of income on bread and grain products; 4.4 per cent on meat; 2.5 per cent on milk, cheese, and eggs; 1.3 per cent on sugar, jam, honey, syrup beverages, chocolate, and sweets; 1.2 per cent on vegetables; and 1.2 per cent on fruit.[40] This shows an increase in the amount spent on meat, sugar, and sweets, which seems to indicate that these foods have become a much more important factor in the diet, a relatively smaller amount of money being spent on vegetables and fruit.

Food did not become more easily accessible until the purchasing power of the Slovene population rose in the last three decades of the twentieth century. With it came an increased consumption of certain staples such as coffee, sugar, and meat. Due to the previously-mentioned periodical unavailability of certain foods in the past, an increased consumption of these items finally stabilized after Slovenia seceded from Yugoslavia in 1991. Shopping for food in large, modern hypermarkets well-stocked with numerous goods became a favourite past-time of many a Slovene, even on Sundays. Perhaps this excellent supply of a large variety of foods and their relatively low prices speak of Slovenia's openness and of its liberal economy.

References

Adrić, I. *Leksikon YU mitologije* [Lexicon of Yugoslav Mythology], Beograd, Zagreb, 2004.

Cvirn, J., Vidic, M. and Brenk, L., *Ilustrirana zgodovina Slovencev* [Illustrated History of Slovenia], Ljubljana, 2000.

Fridl, J., Kladnik, D., Orožen Adamič, M. and Perko, D. (eds) *Geografski atlas Slovenije* [Geographical Atlas of Slovenia], Ljubljana, 1998.

Godina, M. *Iz mariborskih predmestij* [From Maribor's Suburbs], Maribor, 1992.

Godina Golija, M. *Prehrana v Mariboru v dvajsetih in tridesetih letih 20. stoletja* [Food Culture in Maribor in the 1920s and 1930s], Maribor, 1996.

Godina Golija, M. *Prehranski pojmovnik za mlade* [Concepts of Food Culture for Young People], Maribor, 2006.

Ledinek, Š. *Potepanja po poteh Šavrinke Marije* [Roaming the Routes of the Šavrinka Marija], Ljubljana, 2000.

Ložar, R. *Narodopisje Slovencev 1* [Ethnography of Slovene 1], Ljubljana, 1944.

---

38   *Statistični godišnjak FNRJ* 1954: 311.
39   *Statistični godišnjak* 1973: 317.
40   *Statistični letopis Republike Slovenije* 2004: 271.

Mlakar Adamič, J. *Teknilo nam je!* [It was to our Taste!], Trbovlje, 2004.

Reisman, A. *Iz življenja med vojno* [About Life in the War], Maribor, 1939.

Sketelj, P. *Več od srebra in zlata nam sadno drevje da...*[More than Gold and Silver ...], Ljubljana, 1998.

Statistični godišnjak FNRJ 1954 [Statistical Yearbook of the Federal National Republic of Jugoslavia], Beograd, 1954.

Statistični godišnjak 1974 [Statistical Yearbook, 1974], Beograd, 1974.

Statistični letopis Republike Slovenije [Statistical Yearbook of the Republic of Slovenia], Ljubljana, 2004.

Židov, N. *Ljubljanski živilski trg* [Ljubljana's Market Place], Ljubljana, 1994.

Chapter 5

The Stop-Go Era: Restoring Food Choice in Britain after World War II

Derek J. Oddy

The End of the Prewar Diet

In 1939, Britain was an advanced industrial economy with a complex food chain in which raw materials were extensively modified by food processors and refiners. During the Second World War consumers were forced to adapt to a more restricted diet, similar to that of the nineteenth century, as many branded and manufactured foods disappeared from the shops. To appreciate the extent of this change, it is necessary to understand how the food market of the 1930s was being transformed by food manufacturers, retailers and advertisers, all eagerly seeking to influence food purchases. Food consumption was differentiated by income.[1] Apart from low-income families, such as the unemployed, who were unable to participate in the mainstream market, consumers were eagerly buying the new 'ready-to-eat' American products like Kellogg's or Quaker Oats' breakfast cereals, factory-produced biscuits[2] and confectionery and the growing range of tinned fish, milks, cream, vegetables and tropical fruits.[3] Even more fashionable were the ice creams and chocolates available at cinemas and the milk bars and soda-fountains to be found in large towns. The spread of these novelties was promoted by food manufacturers, such as Wall's, who rented refrigerated cabinets to shopkeepers; by the end of the 1930s ice cream was available widely at small 'corner shops' along with tobacco, confectionery and newspapers. Apart from those districts sunk in the poverty of unemployment, even small food retailers provided a very different range of goods in the 1930s than had been available at the end of the First World War:

> almost every kind of domestic and foreign fruit, meat, game, fish, vegetable could be bought, even in country groceries. Foodstuffs that needed no tin-opener were

---

1 Consumers were certainly not homogeneous: Sir John Boyd Orr divided prewar Britain into one-third well fed, one-third adequately fed and one-third whose diet was deficient. Orr and Lubbock 1940: 61.

2 See Collins 1976: 33–6

3 See Reader 1976: 70–1.

also gradually standardized: eggs, milk, and butter were graded and guaranteed and greengrocers began selling branded oranges and bananas.[4]

The Americanization of the British food market was accentuated by widespread advertising in cinemas and the broadcasts of commercial radio stations. In the 1930s Radio Luxemburg was widely available in Britain, bringing advertisements for confectionery and bedtime drinks, such as Ovaltine, to British households.[5] Some advertisements stressed novelty and luxury in an era much influenced by the social behaviour seen at the cinema but, in the age of time-and-motion study satirized by Charlie Chaplin's 'Modern Times', the processed-food market relied on stressing the time-saving advantages of the new products. Young housewives were eager to buy their food materials 'in the nearest possible stage to table-readiness':

> the complicated processes of making custard, caramel, blanc-mange, jelly, and other puddings and sweets, were reduced to a single short operation by the use of prepared powders. Porridge had once been the almost universal middle-class breakfast food. It now no longer took twenty minutes to cook, Quick Quaker Oats reducing the time to two: but even so, cereals in the American style, eaten with milk, began to challenge porridge and bacon and eggs in prosperous homes, and the bread and margarine eaten by the poor.[6]

This dynamic food market came to a sudden end early in 1940 as the wartime economy became formalized. Initially, the restrictions on cargo-space imposed in 1939 limited food imports and the priority given to military food requirements further constrained supplies for the civilian population. The food-rationing scheme which followed in the spring of 1940 was based largely on shipping capacity rather than nutritional advice. It established a tightly controlled regime which lasted until the end of the decade and was characterized by leaving bread and potatoes unrationed, but placed severe limitations on imported foods like sugar and animal products, such as meat and dairy produce, which had not only high status but were also important for their high protein and fat content. Imported fruits disappeared altogether from the diet and fruit and vegetables produced in Britain were only available when in season. It was as though the whole country had been taken back in time to an age when ordinary people ate a diet in which most energy came from starchy carbohydrate foods, while food from animal sources, which provided protein and fat, made only a limited contribution to nutritional status. However, this comparison must not be taken too literally, since Britain's wartime welfare services provided

---

4   Graves and Hodge 1971: 172.

5   Novartis Ltd, the producers of Ovaltine, had the Ovaltineys' club for children. Its main competitor, Cadbury's also sponsored radio programmes and produced a special cocoa tin for children containing miniature animals, backed by its Cococubs' club and magazine.

6   Graves and Hodge 1971: 171–2.

dietary supplements of cod-liver oil, rosehip syrup and concentrated orange juice for pregnant women, infants and young children. By 1941, the fortification of foods to prevent nutritional deficiencies meant that calcium was added to flour[7] and vitamins A and D to margarine. Nevertheless, the high carbohydrate wartime diet was not popular: people found the bread quality unattractive, and disliked the monotony resulting from shortages and the removal of peacetime choice.

A Model for Change

Sir Jack Drummond (1891–1952) formulated a model of dietary change in *The Englishman's Food* (1939) which has relevance to the postwar resumption of choice. It was devised during a time of underconsumption when perhaps as many as one-third of the British population was inadequately fed. Although principally a biochemist, Drummond's knowledge of food consumption in Britain was based on his work in the First World War and the application of vitamin discoveries to the nutrition of the interwar population. This led to his interest in how dietary change occurred:

> The decline in the consumption of bread and flour…is an interesting phenomenon which began to be evident in Great Britain and the United States late in the nineteenth century…It seems to be related to a rising standard of living for the falling curve representing bread and flour is complementary to the rising curve for sugar and sweetmeats. It is also related to a rise in the consumption of meat. These relationships reflect the fact that bread is the staple food of poverty and that people eat much less of it when they can afford to buy meat and indulge in the type of dish with which sugar is eaten.[8]

Drummond's model originated when food imports were freely available and its application to postwar conditions depends upon levels of income, the removal of controls and the return of consumer choice.

The Continuation of Food Controls 1945–1954

The desire for the foods and meals previously enjoyed in peacetime became acute once the war ended. With rationing still in full force, the need to maintain a culture of serving limited portions and not wasting food led to a widespread feeling of deprivation.[9] If people in Britain did not actually feel hungry in the later 1940s,

---

7   Wartime 'National Flour' introduced in April 1942 had an extraction rate of 85 per cent, which made the 'National Loaf' an unappetizing grey colour in the eyes of British consumers.

8   Drummond and Wilbraham 1957: 299.

9   As late as 1948, weekly rations per head were meat: 13 ozs (369 g); cheese: 1.5 ozs (43 g); butter and margarine: 6 ozs (170 g); cooking fat: 1 oz (28 g); sugar: 8 ozs (227 g);

there was certainly a strong desire to eat more – and to eat without the restrictions imposed by rationing. Housewives complained that the smallness of the fat ration made it impossible to cook 'normal' meals in the prewar style. The 'hard currency' crisis,[10] which led to the removal of American dried egg from the shops in February 1946 for several months, accentuated the problems of cooking and devising meals, and the reduction in the size of the loaf, followed by the introduction of bread rationing in July 1946, aroused widespread discontent.[11] Bread rationing remained in force until July 1948, which generated a desire to eat even more bread and flour, provided there was enough butter and preserves to go with the bread, or sugar, fat and dried eggs to bake into cakes or biscuits. Bread consumption rose to 1.87 kg per head per week in 1948 when rationing ended, but while restrictions on the quality of bread-flour remained in force until 1956, home baking continued to be a means of obtaining more satisfaction from what food was available. Consumption of flour peaked at 250 g per head per week in 1954. The general rise in food consumption from the end of the Second World War until food rationing ceased in 1954 is shown in Table 5.1. Cheese consumption recovered from below 60 g per head per week in 1948 to around 80 g per head per week by 1954. The use

Table 5.1   Weekly Food Consumption per Head During Postwar Rationing, 1945–1954

Date	Milk & cream (l)	Cheese (g)	Fruit (g)	Potatoes (kg)	Other Vege-tables (kg)	Bread & cereals (kg)	Sugar (a) (g)	Fish (b) (g)	Fats (c) (g)	Meat (d) (g)
1945	2.52	71	318	1.87	1.16	2.42	414	261	245	746
1946	2.45	72	302	2.00	1.16	2.34	425	299	233	757
1950	2.94	72	513	1.76	1.04	2.32	466	188	329	846
1954	2.89	82	594	1.76	0.89	2.23	599	161	331	955

Source: Department of the Environment, Food and Rural Affairs (DEFRA), *National Food Survey*.

Notes: (a) including preserves; (b) including fish products; (c) Fats = total 'visible' fats, i.e. butter, margarine, lard, and vegetable oils and spreads; (d) Total meat = beef and veal, mutton and lamb, pork, bacon and ham, poultry and sausages.

---

milk: 2pts (1.14 litres); and one egg (57 g). Bread was rationed between July 1946 and July 1948 at 4 lb (1.82 kg) per adult per week and potato consumption was also controlled between November 1947 and April 1948.

10   Sterling was not freely convertible into other currencies after the War, so 'hard' currencies i.e. United States and Canadian dollars, were necessary to buy goods imported from North America.

11   Susan Cooper, 'Snoek Piquante' in Sissons and French 1964: 38, 41. Two-pound loaves (908 g) were reduced by 4 ozs (114 g) and one-pound loaves (454 g) by 2 ozs (57 g).

of eggs doubled to over four eggs per head per week by the mid-1950s and fruit consumption showed a similar increase from around 300 g per head per week in 1946 to 600 g ten years later. Fresh meat consumption fell postwar to a low point of 323 g in 1948–9, after which it rose rapidly. Purchases of meat and meat products recovered generally during the early 1950s, reaching 955 g per head per week by 1954.

Sweets and confectionery were taken off ration in April 1949 but the rush to buy them caused rationing to be reintroduced until February 1953, when its removal was politically timed to occur before the coronation of Queen Elizabeth. Nevertheless, even while the final phase of food restrictions was still in operation, new foods began to enter or re-enter the diet once the 'points system' of food rationing ceased in 1950.[12] Oranges, bananas and other fruit returned to the shops and frozen vegetables began to appear. Although there were initial economic difficulties during the late 1940s, postwar Britain experienced boom conditions and full employment for some twenty years after the war ended. Full employment and wage inflation gave consumers a marked increase in purchasing power especially as the Agriculture Act, 1947, continued the wartime policy of subsidizing basic foodstuffs through fixed prices.[13] This made food relatively cheap, though processed and manufactured foods were not covered in the same way. Even so, the government restricted consumer spending by delaying the repayment of Postwar Credits and limiting the amount of currency that could be spent abroad.[14]

Daily energy intakes in the immediate postwar years are shown in Table 5.2. In 1946, energy intake was actually below wartime levels at just over 2,300 kcals (9,652 kj) per head per day. This was the result of the restrictions imposed on bread and potatoes, which brought the carbohydrate intake down to 305 g per head per day in 1946. In the following year, 1947, fat intake fell to 82 g per head per day,[15] a level which triggered off housewives' protests against the government's continuation of rationing at wartime levels while calling for increased efforts at work to sustain an export drive. The relaxation of food rationing between 1950 and 1954 led to significant gains in energy intakes which rose from 2,474 kcals (10,351 kj) per head to over 2,600 kcals (10,900 kj). These gains resulted principally from the rise in fat intakes to 107 g per day in 1954. Carbohydrate intake, similarly, rose to 340 g over the same period.

---

12   The points system was a method of allocating non-essential foods which were only available spasmodically. It began in 1941. See Burnett 1989: 293–4.

13   For an outline of these events, see Oddy 2003: Chapters 7 and 8.

14   Postwar Credits were a system of compulsory wartime savings on incomes above a certain level. The government delayed their repayment during the postwar sterling crisis.

15   Total fat intake in Table 5.2 includes not only visible fats (butter, margarine, lard and vegetable oils) but also fat contained in foods such as meat, dairy produce and fish.

Table 5.2   The Daily Nutrient Intake in Britain During the Period of Postwar Rationing, 1945–1954

Date	Energy (kj)	Value (kcal)	Protein (g)	Fat (g)	CHO (a) (g)	Calcium (mg)	Iron (mg)	Vit.C (b) (mg)	Vit.A (mg)	Vit.D (µg)
1945	9,937	2,375	76.0	92	309	875	12.7	43	2,908	3.57
1946	9,652	2,307	78.0	86	305	912	14.4	44	2,926	3.43
1950	10,351	2,474	78.0	101	314	1,066	13.6	[43]	3,536	4.30
1954	10,987	2,626	77.0	107	340	1,034	13.4	50	3,911	3.60

Source: DEFRA, *National Food Survey*.

Notes: (a) Carbohydrate (CHO) = available carbohydrate, calculated as monosaccharide; (b) The vitamin C value in brackets is an estimate to include cooking losses.

The postwar increase in energy intake in to over 2,600 kcals per day seems remarkably high, though it reflected people's high energy expenditure at work and at home. The mid-twentieth century industrial economy, ranging from mining, metal production, and manufacturing to transport and distribution, required greater labour inputs than later in the century. The population was also more physically active: many people walked or cycled daily both to work and during leisure activities. In addition, they lived in a chilly domestic environment which relied on a limited use of coal fires with few, if any, labour-saving devices.

The Restoration of Choice

Tables 5.1 and 5.2 show how the consumption of foods was affected by the end of rationing. Bread consumption fell as flour and sugar became available for home-baking, which brought more variety and taste to starchy foods. As the home-baking boom came to an end, flour consumption dropped below 200 g in 1959 and fell steadily thereafter. The consumption of cakes and pastries remained high in the 1950s and 1960s within the range of 150–190 g per head per week, while increased biscuit eating stabilized at around 160–165 g per week from 1959 to 1977. In effect, shop goods became preferred to home-baked foods. The consumption of preserves, which had accompanied the increase in bread consumption, also began a decline which matched the diminishing role of bread in the British diet.

Table 5.2 demonstrates that there was a surge in demand for most of the severely rationed foods as soon as restrictions were lifted. Milk consumption rose to about 2.7–2.9 litres per head per week in the early 1950s, while butter, so scarce during the war, was much in demand with consumption rising to

127 g by 1955 and reaching a peak of 175 g per head per week in 1967. Meat consumption reached over one kilogram per head in the 1960s. The most successful sectors of the meat market were those discriminated against by government feed policies during World War II: pork, bacon and poultry. Then, the size of the bacon ration was a constant source of complaint and fresh pork was scarce. By the late 1940s the consumption of pork, bacon and ham was down to 60 g per head per week. Consumption increased once restrictions were lifted, reaching almost 260 g per week in 1970. Despite the changes in different sectors of the market, overall meat consumption remained remarkably stable: between 1956 and 1989 more than 1kg per head per week was eaten.

Finally, Table 5.3 confirms that the consumption of foods which were unpopular in the Second World War generally declined. Although fresh green vegetables had been heavily promoted by Ministry of Food propaganda, wartime levels of consumption were never maintained once peacetime conditions prevailed in the food markets. During the War, consumption had risen to 500 g per head per week but during the 1950s and early 1960s fell to 400 g. The consumption of fish, frequently stale during the war due to distribution problems and the lack of refrigerated transport, also fell significantly.

Once the immediate postwar shortages disappeared and rationing ceased, consumption of the wartime staples − bread and potatoes − fell from their peak levels of the late 1940s as shown in Tables 5.1 and 5.3. Consumption of bread fell from 1.87 kg per head per week in 1948 to just over 1.5 kg per head per week in 1955. Similarly, potato consumption fell from 2 kg per head per week in 1946 to 1.7 kg in 1955. For both foods this was merely the first phase of a downward trend that continued beyond the end of restrictions. A shortage of sugar had also been a major problem during World War II, particularly when much emphasis was placed upon its importance as a source of energy in the diet.[16] This importance attributed to sugar had given it a special place in Drummond's model of dietary change as the one carbohydrate food the consumption of which would increase with affluence. During the early 1950s, sugar consumption rose rapidly to over 500 g per week in 1955 and remaining above 500 g per head per week until 1963.

---

16   National Archives, CAB74/1. Second meeting of the War Cabinet Ministerial Sub-Committee on Food Control, 1 December 1939. Walter Elliot, Minister of Health, pressed for as big a sugar ration as possible. He wanted 1 lb (454 g) per week but finally accepted 12 ozs (340 g).

**Table 5.3   The Resumption of Choice in the Era of Full Employment:
Food Consumption in Britain per Head per Week, 1950–1970**

(a) Compensation Eating

Date	Bread (kg)	Flour (g)	Cakes and pastries (g)	Biscuits (g)	Sugar (g)
1950	1.64	206	190	104	287
1955	1.56	243	158	145	500
1960	1.29	192	179	161	503
1965	1.15	173	191	165	498
1970	1.08	161	161	163	480

(b) De-Rationed Foods

Date	Whole milk (l)	Eggs (no.)	Butter (g)	Fresh meat (g)	Bacon & ham (g)
1950	2.72	3.5	129	391	128
1955	2.73	4.2	127	517	172
1960	2.75	4.6	161	493	175
1965	2.76	4.8	173	475	179
1970	2.63	4.7	170	450	177

(c) The Wartime Substitute Foods

Date	Fish (g)	Potatoes (kg)	Green vegetables (g)	Other fresh vegetables (g)	Total fresh vegetables (g)
1950	105	1.76	392	433	825
1955	101	1.70	415	415	830
1960	76	1.59	430	427	857
1965	71	1.51	407	406	813
1970	56	1.47	372	394	766

Source: DEFRA, *National Food Survey*.

Table 5.4 indicates that energy intakes began to level off in the 1960s, having reached the highest energy intake of the postwar era in 1963 when 2,650 kcals (11,090 kj) per day was recorded. However protein intake changed little, though it was composed of more animal than vegetable protein by contrast with the wartime diet of more vegetable than animal protein. Protein intake remained stable at around 75 g per head per day until 1968, after which it fell below 70 g per day in 1983. By the 1960s, the nutrient analysis of the diet began to show how accurate Drummond's model of dietary change had been. Carbohydrate intakes were declining, dropping to 343 g per day in 1963, a year in which fat intakes had reached 118 g per day. Carbohydrate intake continued on its downward trend, falling below 300 g per day for the first time in 1973, while fat intake continued to rise, reaching 120 g per head per day in 1969.

**Table 5.4   Daily Nutrient Intake in Great Britain, 1955–1970:
The Return to Free Choice**

Date	Energy	Value	Protein	Fat (a)	CHO (b)	Calcium	Iron	Vit.C	Vit.A	Vit.D
	(kj)	(kcal)	(g)	(g)	(g)	(mg)	(mg)	(mg)	(mg)	(µg)
1955	11,050	2,641	77.0	107	342	1,044	13.5	51	4,199	3.60
1960	11,004	2,630	74.7	115	345	1,040	14.1	52	4,360	3.25
1965	10,837	2,590	75.2	116	332	1,020	13.9	52	4,370	3.13
1970	10,711	2,560	73.7	119	317	1,030	13.4	52	1,350	2.82

Source: DEFRA, *National Food Survey*.

Notes: (a) Fat = total visible fats plus fat contained in other foods (e.g. meat, dairy produce and fish); (b) Carbohydrate (CHO) = available carbohydrate, calculated as monosaccharide.

Food Choice in an Age of Affluence

By the 1970s the age of under-consumption which had characterized the first half of the twentieth century was over, and the era of over-consumption was beginning. Full employment in the formerly depressed areas of the 1930s brought about a levelling-up of living standards. Rising wages meant that food spending fell as a proportion of total household expenditure from 29.8 per cent in 1951 to 20.2 per cent in 1971.[17] The new norms of consumer behaviour were being set by the food industry's advertisements and the displays on the shelves of the new postwar retailing phenomenon, the supermarkets. None of this meant fundamental change in purchases of food: during the 1940s, people in Britain had simply wanted more of the same foods they had eaten before the war and by the 1960s this had

---

17   *Social Trends* 1972: Table 54.

been achieved. Table 5.5 illustrates how longer-term changes in the postwar diet developed. The beginning of this process had been evident from the mid-1950s onwards: bread and cereal products continued to decline and flour consumption also fell, apart from several years in the late 1970s during the time of rapid inflation. From 1965 the consumption of cakes fell below 100 g in 1974 but stabilized at between 70–90 g per head per week for the rest of the twentieth century. Despite Drummond's predictions, sugar consumption also fell to under 400 g in 1973 – regardless of its promotion as a loss-leader by the supermarkets. This downturn is shown in Table 5.5(a). Besides the fall in sugar consumption, Table 5.5(b) shows that overall demand for meat declined slightly in the last quarter of the twentieth century. Falls in consumption of fresh meat and sausages were partially offset by the remarkable increase in poultry consumed, of which chicken was the prime component. In 1950, only 10 g per head per week was eaten but by the late 1990s this figure had risen to 250 g.

Table 5.5(c) provides evidence of the extent to which the 'traditional' British diet of the first three-quarters of the twentieth century was changing. The consumption of breakfast cereals continued to rise from around 40g per week in the early 1950s to more than 100 g per head per week after 1980. The postwar rise in the consumption of liquid whole milk ceased after peaking at 2.8 litres in 1962–3 and was followed by a decline to 2 litres in 1984 and 1.5 litres in 1988. During the 1990s, the consumption of liquid whole milk collapsed from 1.1 litres in 1991 to 0.63 litre in 1999. This change was offset by the rise in skimmed and semi-skimmed milk consumption during the 1980s. Between 1988 and 1994, low-fat milk consumption doubled from 0.54 to 1.07 litres per head per week. Complementing this change was the rise in popularity of yoghurt and fromage frais, which grew from novelty food status in the 1960s and early 1970s to a consumption of over 100 g per head per week from 1993 onwards. During the 1960s, the postwar decline in consumption of fresh green vegetables was arrested, as frozen vegetables such as peas and beans became popular. With an expansion in the variety of frozen products, consumption rose to 200 g per head per week in the 1990s.

Table 5.5   Trends in Food Consumption per Head per Week, 1975–1995

(a) Staple Foods

Date	Bread (g)	Flour (g)	Sugar (g)	Butter (g)	Margarine (g)	Lard (g)	Whole milk (l)	Tea (g)
1975	956	149	319	160	74	55	2.71	62
1980	882	161	317	115	108	51	2.36	58
1985	878	115	238	80	106	41	1.89	49
1990	797	91	171	46	91	23	1.23	43
1995	756	57	136	36	41	13	0.81	39

(b) The Changing Demand for Meat

Date	Beef & veal (g)	Mutton & lamb (g)	Pork, bacon & ham (g)	Poultry (g)	Sausages (g)	Total meat & meat products (g)
1975	238	120	220	160	92	1,054
1980	231	128	266	189	92	1,140
1985	185	93	235	195	84	1,042
1990	149	83	202	226	68	968
1995	121	54	186	237	63	945

Note: Total meat includes poultry, sausages and meat consumed in meat products, such as pies and pasties.

(c) New Foods

Date	Breakfast cereals (g)	Skimmed milk (a) (ml)	Yoghurt (b) (g)	Fresh fruit (g)	Fruit juice (ml)	Frozen vegetables (g)	Oils & fats (c) (g)
1975	87	7	24	495	38	91	27
1980	99	20	48	590	87	130	43
1985	114	244	74	524	148	170	57
1990	127	709	97	605	202	185	97
1995	135	1,103	127	672	244	200	128

Source: DEFRA, *National Food Survey*.

Notes: (a) including semi-skimmed milk; (b) includes fromage frais; (c) including vegetable oil, low fat spreads and reduced fat spreads.

Evaluating Changes

In postwar Britain, the food industry's new products benefited from advances in low-temperature technology and other new scientific discoveries which were applied to food processing.[18] Their impact in changing consumption of the broad categories of food — bread, meat, milk, vegetables — was not immediately obvious, but within each sector significant shifts developed by which one food was substituted for another. This seemed a matter of eating style rather than anything else: for example, Britons continued to consume meat, potatoes and vegetables as their main meal. One commentator on the second half of the twentieth century saw the trend as:

> a switch from traditional roast beef, roast potatoes and cabbage in the middle of the day to microwaved chicken, frozen chips, and courgettes as an evening snack.[19]

The third quarter of the twentieth century was a period when food in Britain was of a predictable quality and safe to eat. These were not conditions conducive to major change in food consumption but rather of continuing to eat and enjoy well known and liked foods. Under conditions of stable prices and safe food, British consumer behaviour followed Drummond's model, both by eating more and by increasing their consumption of the more expensive animal products that contemporary nutritional advice explained was essential for health and growth. This suited the inherently conservative nature of the British consumer and echoed the old nineteenth-century advice to 'Eat your way to good health'. Tampering with the food supply was unwelcome, as considerable opposition to the fluoridation of drinking water, which began in 1956, showed.[20] The year 1956 also saw two other events which were important for the future of food consumption: first, the final wartime food regulation was removed and national flour ceased to exist; second, commercial television began and advertisements for food were brought into people's living rooms. The beginnings of a more sedentary lifestyle could be said to start from this point.

From the mid-1960s Britain entered an economic transition which eroded the stability of the 1950s. Unemployment began to rise, until one million people were out of work by 1967. An era of rapid inflation began in the early 1970s with the decimalization of the currency and entry into the Common Market destabilizing food prices. It was accentuated by the oil crises of 1973 and 1979, so that the 1970s

---

18   See Pyke 1970: Chapter 9, for developments in canning, quick-freezing, accelerated freeze-drying, dehydration, the use of radio-activity (irradiation) and chemical additives.

19   *Fifty Years of the National Food* Survey: Chapter 5, 'The British diet from 1940 to 1989' [by David H. Buss].

20   Oddy 2003: 210. The retention of the policy of fortification in the case of some foods also aroused some consumer concern.

experienced the transition to a period of expensive food. The immediate reaction amongst consumers was to reverse purchasing patterns in favour of cheaper filling foods, such as bread, potatoes, rice and margarine. The food industry reacted by introducing cost-cutting techniques in its ready-prepared dishes. Meat content was reduced and its quality fell, as mechanically recovered meat (MRM) began to be used, with food manufacturers concentrating on the value-added component of ready-prepared food dishes.

By the 1970s, the sedentary life style associated with widespread car ownership, limited physical activity, eating in front of the television set, and snacking rather than eating formal meals had become commonplace. Table 5.6 shows that, although energy intakes began to decline from the mid-1960s onwards, the proportion of fat in the diet increased: its association with weight gain and its effect on people's health became linked to the rise in cardiovascular disease.[21] Further threats to the consumer, which began to modify the choice of foods, arose from questions of food safety.[22] From the 1960s onwards the food market became a battleground between the food processors and refiners, as the dairy industry (assisted by the dental profession's concern regarding the state of children's teeth) blamed high sugar consumption for Britain's health problems and the margarine producers blamed the dairy industry for supplying consumers with saturated fats. Sugar consumption fell only gradually despite the authoritative assault by Professor John Yudkin.[23] Consumption of meat, fish, eggs and fats all reflected some modifications in consumer behaviour but concern that diet could be a factor in coronary heart disease affected fat consumption more than sugar. Table 5.6 shows that fat, as a source of energy, began to decline from its peak in 1985 while the intake of carbohydrate reversed its previous downward trend.

21 See Oddy 2003: 210, for the even earlier rise in coronary heart disease between 1948 and 1958 and pages 220–23 for the rise in obesity in Britain and the increasing incidence of Type 2 diabetes during the 1990s.

22 Oddy 2003: 231, for the impact of Rachel Carson's *Silent Spring* in Britain from 1965 onwards. It drew attention to the use of chemicals in crop production, stimulated fears about the use of chemicals to promote animal growth, and led to the beginnings of the fears that food additives, especially dyes, might be carcinogens.

23 Yudkin 1972.

Table 5.6   Percentage Source of Daily Energy Intakes in the British Diet

Date	Energy value (kcal)	Protein (%)	Fat (%)	CHO (%)
1946	2,307	13.5	33.6	52.9
1950	2,474	12.6	36.7	50.7
1955	2,641	11.6	36.6	51.7
1960	2,630	11.4	39.3	49.3
1965	2,590	11.6	40.4	47.9
1970	2,560	11.5	41.8	46.5
1975	2,290	12.6	42.2	45.2
1980	2,230	13.0	42.6	44.4
1985	2,020	13.3	42.6	44.1
1990	1,870	13.5	41.6	44.9
1995	1,780	14.2	39.8	46.0

Source: DEFRA, *National Food Survey.*

Note: The National Food Survey did not take account of food or drink eaten outside the home before 1995. For the inclusion by the National Food Survey of food eaten outside the home and its effect on energy intakes, see Oddy, 2003, ICREFH VII, 310.

From the late 1960s onwards, consumers became confused about the relationship between food and health. Foods which traditionally were of high status for their protective qualities, such as meat and dairy produce, came under attack as competing branches of the food industry mounted campaigns against each other's products, while weight-reduction programmes featured continually in newspapers and magazines. These were seldom compatible with food manufacturers' advertisements to consume more, particularly as manufacturers were resistant to food labelling. Consumers' apprehensions about food grew throughout the final third of the twentieth century, during which Britain experienced a succession of safety problems, including typhoid in imported canned meat, salmonella, *E.coli*, foot-and-mouth disease, Bovine Spongiform Encephalopathy (BSE) and its human variant Creutzfeld Jacob Disease (CJD). Food historians should recognize that these dangers to health, which were associated with a greater processing of food, developed as food manufacturers' influence over the food supply became more extensive. The food industry's emphasis on the need for less cooking by the provision of more prepared dishes and complete meals ready for oven or microwave suited the snacking and fast-food habits which were being adopted in the late twentieth century. By urging the use of their products, food manufacturers and processors were encouraging consumers to ignore raw food materials and traditional cooking in favour of processed foods and prepared dishes. Such encouragements differed little from the time-saving advertisements of the 1930s.

However, in the last twenty years, British food manufacturers have also faced the challenge of television celebrity chefs urging viewers to try their hands at real cooking again. Not to be outdone, manufacturers and supermarkets began featuring celebrity chefs in their products and distribution systems.[24] Celebrity sauces and cooking supplements (Figure 5.1), direct association between chef and shop and identification of television ingredients on the retailers' shelves retained the commercial control over consumers' kitchens and limited independent purchasing and cooking still further.

"Have you tried this? It used to be all the rage"

Figure 5.1 'Have you tried this? It used to be all the rage'

Source: *Private Eye* 16 March 2007, 7, with permission.

---

24   As, for example, the association between Jamie Oliver and Sainsbury's food advertisements.

References

Agriculture, Fisheries and Food, Ministry of *Household Food Consumption* (annual series), later as Environment, Food and Rural Affairs, Department of *National Food Survey*, London.

Agriculture, Fisheries and Food, Ministry of *Fifty Years of the National Food Survey 1940–1990* (ed. J.M. Slater), London, 1991.

Burnett, J. *Plenty and Want: A Social History of Diet in England from 1815 to the Present Day*, London, Third edition, 1989.

Central Statistical Office, *Social Trends, No.3 1972*, London, 1972.

Collins, E.J.T. 'The "consumer revolution" and the growth of factory foods: changing patterns of bread and cereal-eating in Britain in the twentieth century', in Oddy and Miller (eds) 33–6.

Drummond, J.C. and Wilbraham, A. *The Englishman's Food*, London, revised edition, 1957.

Graves, R. and Hodge A. *The Long Week-end*, Harmondsworth, 1971.

Jacobs, M. and Scholliers P. (eds), *Eating Out in Europe: Picnics, Gourmet Dining and Snacks since the Late Eighteenth Century*, (ICREFH VII) Oxford, 2003.

Oddy, D.J. and Miller, D.S., *The Making of the Modern British Diet*, London, 1976.

Oddy, D.J., *From Plain Fare to Fusion Food: British Diet from the 1890s to the 1990s*, Woodbridge, 2003.

Orr, J.B. and Lubbock, D., *Feeding the People in Wartime*, London, 1940.

Pyke, M. *Food Science and Technology*, London, Third edition, 1970.

Reader, W.J. *Metal Box: A History*, London, 1976.

Sissons, M. and French, P. (eds), *Age of Austerity 1945–1951*, Harmondsworth, 1964.

Yudkin, J. *Pure White and Deadly: The Problem of Sugar*, London, 1972.

PART 2
Industrial and Commercial Influences on Food Consumption

Chapter 6

How Food Products Gained an Individual 'Face': Trademarks as a Medium of Advertising in the Growing Modern Market Economy in Germany

Hans Jürgen Teuteberg

Today nearly half of the German population is estimated to be overweight and 20–25 per cent are afflicted with pathological obesity. This epidemic of corpulence is so significant that we can reasonably call it a 'new social question'.[1] The drumfire of modern advertising is said to be one of the causes of an excessive, unhealthy consumption of food, which in pre-industrial times was restricted to the small circle of the upper class. The aim of this chapter is to describe the early rise of modern advertising and to focus on a number of leading trademarks for foodstuffs and luxuries during the decades around 1900, to gauge the influence they had on variations in daily nutrition in Germany and to compare this with the period of mass consumption in the late twentieth century.

As everybody knows, many foodstuffs bear specific trademarks. These labels form a bridge between producers and consumers, and they give certain products a lift that differentiates them from the mass of goods. Trademarks have four main functions. First, they are signs of non-interchangeable identities and arouse anticipations and memories of particular consumption habits and practices that are helpful for consumers when searching the markets. Second, branded articles make regional and qualitative differences more transparent, as well as price levels. Third, they promote sales by establishing certain standards of quality and quantity, and sometimes also of price. Finally, they give protection to consumers and, at the same time, proclaim the proprietorship of special products. Their socio-cultural relevance is, of course, still greater.[2]

---

1   Neuloh and Teuteberg 1979.
2   Hellmann 2003.

The Rise, Differentiation and Shaping of the First Modern Trademarks, 1850–1914

The practice of giving commodities distinguishing signs has many precursors in history but the acceleration in the use of trademarks in Germany is limited to the period from the middle of the nineteenth century onwards, when rapidly expanding urban populations and the eclipse of traditional agrarian self-provision made it necessary to look for new systems of supply.[3] Urban consumers, along with factory owners and wholesale traders, felt an urgent need for more market information.

Those foodstuffs traditionally sold to households in loose form, without any packaging, were now freed from their shells, skins, husks and other peelings. They might be preserved and were often divided into smaller portions and combined with ingredients and seasonings. They were also put into attractive wrapping that could be used for printing messages from the producers to the consumers. The development of trademarks thereby received a new impetus. Most striking at first was the role of newspapers.[4] After the mid-1860s it became common to place small advertisements, not only to give information about the address of the producer, package size and price, but also to establish a sign that could easily be remembered. It was recognized that messages must have utility but also encourage the desire for buying. These advertisements gradually changed their appearance through the employment of professional graphics. Attention was raised by the use of pointing fingers, eye-catching fonts and edges, and short slogans. A leading economist criticized this new form of advertisement as inhibited, with insufficient real information for the buyer.[5] But in the 1870s this was addressed with the emergence, first of special workshops for professional graphics, and then 'advertising bureaux' which brought a more aesthetic and informative approach.

The nineteenth century also witnessed the birth of the advertising poster.[6] The first technical steps came with the invention of lithography around 1800. The extensive use of posters during the revolution of 1848/49 then brought official regulation. The printer Ernst Litfass in Berlin was allowed to set up 150 advertising pillars *(Litfass-Säulen)*, like those he had seen in London and Paris. This first 'king of advertisements' found many imitators, so that by 1900, 52,000 Litfass-pillars had been erected in German cities. Apart from the street, there were many advertisement posters in the big new stores. From the 1890s, simple messages about quantities and prices were enhanced by the use of colour and artistic designs to combine text and illustration in a more meaningful way. The

---

3   Buchli 1962/3, Hamlin 1923, Meldau 1967, Hillebrand 1973, Gombrich 1982.

4   See for the following descriptions and the whole chapter, Reinhardt 1993, Borscheid and Wischermann 1995. Wischermann 2000.

5   Sombart 1902.

6   Schindler 1972, Hollmann 1982, Reichwein 1980, Schwarz 1990.

new photolithography allowed the production of much larger posters.[7] Again, the 1890s represented a threshold in the shaping of announcements, packaging protocols, letter-heads and catalogues. Short headlines became fashionable, or banners of only one word with symbolic illustrations.

Because all announcements printed on paper naturally had a transitory character, the sheet-metal industry started to produce large enamel plates as a new form of advertisement.[8] These flat coloured signboards were like posters, with very short messages concerning one product, primarily with special reference to food and to luxuries. Examples include the meat soup of the Swiss firm Maggi or the medicated Italian wine Martini. Besides announcements, posters and enamel-plates, shop-window displays and paintings on gable walls, the idea spread of using illuminated letters at night and the advertising potential of the sides of buses and other wheeled vehicles.

By the turn of the century the several marketing functions of advertising were already clearly recognized. An announcement in a newspaper, journal, or, later, in the cinema, required the customer to pay attention. A poster, on the other hand, had a subliminal influence, sometimes against the consumer's will. Normal 'ads' on paper intruded directly into the domestic sphere and broadcasting after 1920 and, later, television took this further with their considerable technical advantages.

Examples of the First Successful Advertising of Food Trademarks

The pioneers of German poster art before the First World War were artistic leaders in this medium. Examples include the laughing innkeeper with a bottle of *Dornkaat* (a hard liquor), the Munich waitress with a glass of *Spatenbräu* (a famous Bavarian beer), and the red circle around the heart of the decaffeinated *Kaffee Hag,* a Bremen coffee that is still sold today. Advertisements in journals profited especially from this artistic revolution.[9] Following foreign examples, certain food trademarks prevailed such as *Liebig's Extract of Meat, Kathreiner's Malzkaffee, Knorr's Suppenwürfel* and Bahlsen's *Leibniz Cakes.* Also famous was the 'bright head' or 'bright mind' (*hellen Kopf*) symbol printed on the small baking-powder bags mass produced by August (Dr) Oetker. If we look at the development of the illustrated advertisements of a famous Swiss chocolate company in two German family magazines between 1860 and 1885, we can see that what at first were long texts were eventually condensed to short headlines: illustrations became much more important, and an emphasis upon cheerful emotions gradually replaced information about foodstuff technology, physiological quality, or even price and

---

7 Müller-Brockmann 1989.
8 Riepenhausen 1979, Feuerhorst 1980.
9 Reuleaux 1884, Naumann 1908, Leitherer 1987.

taste.[10] The purpose was clearly to produce a positive stimulus for what at that time was a relatively expensive luxury.

This changing content of advertising had to do with the emerging influence of psychology. Appeals were increasingly made to Christmas emotions, the love of animals, the large field of human vanity and, to a small extent, sexuality. But health problems were never touched upon. The main points of emphasis were already trusted socio-cultural images and these were used to push the customer's demand in a certain direction. All of these messages referred to optimistic wishful-thinking and what was considered to be the spirit of the age: the idea of general human progress. Advertising for the sake of sales promotion was closely linked to mass production, with a view to making goods with particular trademarks available with consistent quality and price everywhere. Sometimes this led to controversy, for instance, when the margarine industry tried, through advertising, to suggest that its new product had the same quality as farmers' 'best butter'.[11]

Overall we can say that the foundations of modern German advertising, and specifically for trademarks, were laid in the decades around 1900. Between 1894 and 1913, after the enactment of protection for labels, the number of these registered signs for trademarks at the Imperial Patent Office in Berlin increased threefold and in the last year before the First World War achieved about 320,000 registrations (Table 6.1).[12]

Table 6.1   New Registrations for Trademarks in Germany, 1894–1913

Year	Registrations	Year	Registrations	Year	Registrations	Year	Registrations
1894	10,781	1899	9,761	1904	15,297	1909	23,371
1895	10,736	1900	9,727	1905	16,564	1910	25,963
1896	10,882	1901	9,924	1906	17,872	1911	26,602
1987	10,477	1902	11,168	1907	18,615	1912	29,507
1898	10,638	1903	12,482	1908	20,098	1913	32,115

Source: Reinhard 1993, 436.

It is interesting to note that 74,000 trademarks were allotted to foodstuffs and luxuries, but only 14,000 to medicaments and 13,000 to cosmetics. The rapid increase of trademarks in Germany, which gave certain goods a name with steady attributes and a certain guaranteed quality, was indicative of the growing needs of an individual lifestyle and of emancipation from pre-modern, social hierarchical ideas of value.

---

10   Schlegel-Mathies 1987.
11   Schlegel-Matthies 1987: 302.
12   Kaiserliches Patentamt 1893ff, Cohn 1902.

One of the main ideas of labelling was on the one hand to protect the consumer against fraud and adulteration and, on the other hand, to give honest producers and tradesmen greater confidence in the market and protection from imitations. The Imperial Patent Office in Berlin started a register in 1874 to prevent errors and duplications, and this contained exact descriptions of all goods and their trademarks. As signs of distinction, these marks could be attached not only to the registered goods, but also to packaging, price lists and business letters. An entry in the register gave protection against imported products which could be confiscated by the state if they violated the rights of German trademark owners. There was also legislation directed against unfair advertising and enforcing copyright for patterns and models. But for the courts it was often not easy to separate overstated advertising from low gossip.

Turning now to the period from 1945 to the present day will make it possible for us to compare the first modest entrance into the modern consumer society during the decades around 1900 with the mass consumption of the late twentieth century.

From 'Black Market' to Internet Commerce

The first ten years after the collapse of Hitler's Third Reich in May 1945 saw a transition from the deepest need amongst the bombed-out ruins to a modest form of normal life.[13] The most basic of daily foodstuffs were the first priority but, because a centrally controlled economy with rationing remained for four years, the black market extended its range. With barter and illegal business so common, the very idea of advertisements seemed absurd. The black market spectre ended on 21 June 1948 with the reform of the old currency. Instantly Germans could now again buy some of the things that they had only dreamed of before. But the general economic recovery remained rather slow at first, and the first concern was for most urban households to achieve sufficient nutrition. The limited financial means of ordinary families were concentrated on what amounted to an 'eating wave' (*Fresswelle*).[14] Well-known food companies started to bring out new products and advertising campaigns. Early examples were the margarine *Sanella* and the Maica meat preserves *Sülzkotelett* (jellied chop) and *Eisbein in Aspik* (pickled pork). By 1952, West German men were already, on average, two kilograms overweight, and in 1954/55 consumption per head reached 3,000 kcals [12,500 kj] per day. The so-called 'welfare belly' (*Wohlstandsbauch*) was the first sign of economic growth, and corpulence was taken as a symbol of the country's steady recovery.

The change to a free market in the 1950s was accompanied, rather like in 1920s, with a tendency to American-style advertising. This was encouraged by the use of products from the United States, such as the cigarette brand Lucky Strike,

---

13   Schmidt 2007.
14   Prinz 2003, Pierenkemper 2007.

as a form of currency on the black market. Also the 'CARE parcels' sent in the first difficult years after 1945 helped to position American goods for the coming upsurge in consumption.[15] Coca-Cola became a symbol of the American quality of life. A poster showed the Statue of Liberty in New York Harbour holding a bottle of Coca-Cola where the torch should be, and a headline announced this as a 'symbol of friendship'. Later German Trade Unions proclaimed the need for a shortening of working hours and Coca-Cola produced a poster with the slogan 'the pause that refreshes' (*Mach mal Pause*). So this American global enterprise gave the Germans a new optimism that they could return to the fold of Western nations. Not far behind came Hawaiian toast, French fries, tomato ketchup and milk shakes, but also ice-cream, pineapple chunks, and frozen vegetables, all of which were understood as characteristic of the process of Americanization.

All this was closely connected with the spread of self-service supermarkets, synthetic packaging and the electrification of the kitchen.[16] A poll in 1955 indicated that modern technology was increasingly a part of taken-for-granted living standards. The domestic refrigerator was owned by 40 per cent rate of households and finally, by 1970, it had made its transition from an object with high social prestige to an every-day piece of kitchen equipment. This social trickle-down affected many other commodities in Germany during this period.

It is also interesting to see that 1950s posters became more aesthetically pleasing as the advertisements of the Stollwerck chocolate company from Cologne showed. West Germans had clearly joined the welfare society by about 1960 and those posters that had remained somewhat functional in design were replaced by full-colour printing. This was partly to reach those German youngsters who were influenced by the fashions of Swinging London's Carnaby Street, including that cheerful, mini-skirted icon, Twiggy. Slimming was revived as a social ideal, similar to the ideas of the health reform movement around 1900. Even children were now chosen as bearers of sales messages. A good example for this was the former cook from Munich, Friedrich Jahn, who founded roast-chicken restaurants (*Backhendl Stationen*) in 140 cities in Europe and the United States under the name *Wiener Wald* (Vienna forest). Many of his posters included mothers and children with a headline such as: 'On Sunday the kitchen will stay cold – and we will go to the *Wiener Wald*.'

Large enterprises, with a growing public-relations staff, began to inform their customers in a more realistic and factual way about their goods in order to win general confidence. This increased care for image was the result of new motivational research which required a new advertising language. This was the time when the American expression 'marketing' as part of a science of industrial management, was adopted everywhere in Germany. As the American sociologist Peter Drucker at the time identified, there was an increasing need for the coordination of present and future production through a combination of research on

---

15   Kriegeskorte 1992.
16   Lummel ICREFH IX, 2007.

markets and consumption, sales technology and advertising. The advocates of this approach to sales promotion were recruited by all the large firms and their socio-psychological methods were used in long-term sales strategies and advertisement campaigns.[17] The shaping of products and advertisements were for the first time generally linked. As in the 1950s, trends in popular culture were driven by ideas from the USA: long hair, blue jeans and hippy clothing, naked bosoms of the Flower Power era, and Easy Rider motorbike gangs, but also new departures in eating and drinking habits, such as the increasing adoption of vegetarianism and, from the 1980s, the demand for organic foods.

A longer lasting influence on advertising styles in the 1960s came from Pop Art.[18] An example was the cutting of photographs from advertising catalogues which were then put together in a style that was inspired by surrealism. Advertising posters which followed this trend, the comic strip approach of Roy Lichtenstein or Andy Warhol's silkscreen printing, were often imitated. The North-German beverage firm Sinalco, for instance, used the comic strip style for many years for the promotion of its lemonades, addressing younger consumers with an international sensibility. It became possible for draughtsmen to earn a lot of money with imitations of American comics. Victor von Bülow (pseud. Loriot) drew at first anonymously for a leading German liquor enterprise but later achieved fame as a television personality and book author. He created caricatures based on the eating and drinking habits of the late twentieth century. Another caricature success was the nervous little man named *Bruno*, used by a big cigarette company to show human mistakes and how smoking can be a relief. This promotion indicated that television was becoming a serious competitor to printed announcements, posters and cinema advertisements.

In recent decades there has been greater differentiation in advertising but also a return to some classic approaches. In the latter category it is striking that the names of well-known personages have once more been used as sympathetic bearers of advertising messages. In the 1950s and 1960s film stars were used in Germany, as in the USA. In the nineteenth century it had been famous men such as Napoleon or Goethe or opera singers who were used in advertisements as a form of trademark, and after the Second World War the pictures of sports heroes or television entertainers were also a way of catching the attention of customers. Often the accompanying message was that consuming a special food or drink product was a means of achieving a higher social prestige.

The 1990s saw many structural changes in a short time. The overall significance of the advertisement as transmitter for mass communication can be demonstrated with some figures. In one year, 1999, the enterprises of the re-united Germany spent DM61 billion on advertising, not only on a great variety of insertions in newspapers and journals, and posters, but also on the broadcast media. In 1994, the first private radio and television station RTL (*Radio et Télévision Luxembourg*)

---

17 Hellmann 2003: 114–62, 277–344, Drucker 1998: 149–58.
18 Spitzer 1962.

made a profit of DM2.8 billion on its advertising alone, reaching 830,000 German households.

The 1960s saw the stirrings in Europe and the USA of a consumer movement critical of the whole system of a capitalist consumer society. For a while this became mainstream, especially among students and other groups of urban youth from the middle classes. The food market was not really touched at first by these consumption rebels and their myth of a counter–culture, but the Green Movement which followed achieved a political status, and created an alternative set of market dynamics of its own. It called for the consumption of organic food (*Biokost)* which has both health and environmental implications. Leading corporations such as Nestlé, Unilever, Kraft Foods and Danone all sought to exploit this new opportunity with their trademark products. Since the 1990s this sector has grown in importance alongside traditional themes such as taste and convenience.

More recently, the growth of the internet has brought into being 'electronic commerce'.[19] In a few years this has changed from an opportunity for computer nerds to play online games to a huge mass market. In 1999 around a quarter of German trademark users had already bought something in this way. Between 1998 and 2000 the number of personal computers with internet connections grew rapidly, as did the number of websites. The online sale of goods and services increased in these two years from DM30.4 billion to DM94.3 billion, with trademarks playing a role in these new communication channels.

But it is also true that that there are limits to this new sales phenomenon. Successful marketing, for example, is best achieved when the goods are not too complicated and do not require expert advice, just as with a mail-order business. Typical are the downloading of music, video-clips and the booking of tickets for travelling and simple bank services, but it is doubtful whether e-commerce can supersede all traditional bearers of advertisements in the food trade. A likely result, as in the past in this sector, is greater differentiation. The classical carriers of advertising will survive, as in previous communication innovations. The importance of trademarks will be underlined by the durability and contents of foodstuffs, and by online prices. But to order fresh foodstuffs through the internet and guarantee quick, direct transport from the producer to the customer is still problematic.

Conclusion

As has been shown, branded articles as a special form of advertisement have not only employed textual messages but also symbols and pictures, often combined with slogans. The contribution of modern advertisements to healthy eating has always been limited because the target of the food industry is always to increase sales. The early influence of psychology and then marketing strategies meant that

---

19   Teuteberg 1998: 294–409.

appeals were made not so much to price, technical or bio-chemical details, but primarily to taste and to emotional triggers such as memories of the past, future dream worlds, vanity or sexuality, and the search for greater social prestige. The numerous links between nutrition and health, with their often negative implications, remained an object of public debate and state regulations, but they did not mesh with the new, optimistic tenor of advertisements for branded food products. From the 1980s a few food manufacturers began to point out the health benefits of their products, for instance the margarine industry in the early stages of the debate about the danger of obesity. The food industry has developed new products and then sold them according to contemporary fashions for more than hundred years. This account has tried to relate industrial food innovations to the changing spirit of the time and to ask what were the deeper meanings behind these food trademarks. It demonstrates clearly that the advantages of modern advertising had already been grasped in Germany before 1900.

The increasing numbers of trademarks were not seen as 'hidden persuaders' of powerless and passive consumers, as the US author Vance Packard maintained in 1958.[20] On the contrary, customers identified an opportunity to benefit through their buying power from the competition between producers, while at the same time protecting themselves against the adulteration of the goods, which in pre-industrial times had always been a tremendous problem. But these new waves of advertising were subject to resource constraints and they always had to follow the continuously changing dynamics of lifestyle.

There can be also no doubt that leading foodstuffs and luxuries and their brand names played an extraordinary role in the development of advertising, because in the rising flood of products they had the function of lighthouses. Since the late nineteenth century trademarks have been orientation guides and focal points for the breakthrough to modern mass consumption. They were concrete indications of the birth of numerous technical, economic and social innovations and reflected the increasing variety of foodstuffs. And last but not least, they encouraged trust in special food qualities.

Champagne, chocolate/cacao and preserved meat soup from America were the first products to undergo price falls, which changed their status from being expensive goods to the objects of modern marketing. They were followed by coffee substitutes, margarine, sweets, beer, sausages, sea food and some tropical fruits. These price reductions were evidence supporting the thesis of the sociologist Norbert Elias that formerly very rare luxuries, which had been symbols of social prestige, in the course of modernization always trickle down from the top of the consumption pyramid to the level of the not so well-off social classes.[21] But there are also some contradictions, as the growth in margarine consumption demonstrates. Both directions of integration reflect the growing variety of foods together with the increasing individualization of the process of consumption. The

---

20   Packard 1958.
21   Elias 1969.

growing range of common victuals, beverages and luxuries during the last 150 years has been limited to only a few examples of this fundamental change in food history. A first dip into some of the poster collections in Germany shows that, at a minimum, there were more than two dozen sorts of foodstuffs and luxuries integrated into the innovative circle of advertisements around 1900. Further research is required to analyse the ways in which each of them was received. This outline has concentrated on Germany, but it can be supposed that in other European economies the modernization of communication channels proceeded in a similar way, although at a different pace and in varied phases.

The impact of food trademarks on changing eating and drinking habits, like food advertisements in general, cannot be quantitatively measured. But this chapter indicates that a complete food history requires not only the traditional discussion of market prices, household budgets, consumption per head and calories, or narrative subjective reports about meals, but also a consideration of the changing structural influences of the advertising media on the contemporary emotions of consumers.

References

Borscheid, P. and Wischermann, C. (eds), *Bilderwelt des Alltags. Werbung in der Konsumgeschichte des 19. und 20. Jahrhunderts. Festschrift für Hans Jürgen Teuteberg* [Daily Life in the World of Illustrations. Advertising in the Consumer Society], Stuttgart, 1995.

Buchli, H. *6000 Jahre Werbung. Geschichte der Wirtschaftswerbung und Propaganda* [6000 Years of Advertising. An Economic History of Advertising and Propaganda], 3 vols, Bern 1962–63.

Drucker, P.F. 'The Discipline of Innovation', *Harvard Business Review* 76, 1, 1998, 149–57.

Elias, N. *Über den Prozeß der Zilisation. Soziogenetische und psychogenetische Untersuchungen* [On the Process of Civilization: Sociogenetic and Psychogenetic Inquiries], vol. 2, Bern, 1969.

Feuerhorst, U. *Email-Plakate. Eine Auswahl von 140 Email-Plakaten mit einer Einleitung und Geschichte* [Email posters. A Selection of 140 Email Posters with an Introduction and History], Dortmund, 1980.

Gombrich, E. *Ornament und Kunst. Schmuckbetrieb und Ordnungssinn in der Psychologie des dekorativen Schaffens* [Ornament and Art. Jewelry-Making and a Sense of Order in the Psychology of Decorative Creativity], Stuttgart, 1982.

Hamlin, A.D.F. *History of Ornament: Renaissance and Modern*, New York, 1923.

Hartog, A.P. den 'The Role of Nutrition in Food Advertisements: The Case of the Netherlands', in ICREFH III, 36–50.

Hellmann, K.U. *Soziologie der Marke* [A Sociology of Brands], Frankfurt, 2003.

Hillebrand, H. (ed.), *Graphic Designers in Europe*, 4 vols, Fribourg, 1973.

Hollmann, H. (ed.), *Das frühe Plakat in Europa und den USA, 3: Deutschland* [The early Poster in Europe and the United States], Berlin, 1980.

Kaiserliches Patentamt (ed.), *Warenzeichenblatt* [The Trademark Gazette], Berlin, 1893–1994.

Kriegeskorte, M. *Werbung in Deutschland, 1945–1965. Die Nachkriegszeit im Spiegel von Anzeigen* [Advertising in Germany, 1945–1965. The Postwar Period in the Mirror of Advertisements], Köln, 1992.

Leitherer, E. and Wichmann, H. *Reiz und Hülle. Gestaltete Warenverpackungen des 19. und 29. Jahrhunderts* [Designed Product Packaging of the 19th and 20th Century], Basel, 1987.

Lummel, P. 'Born-in-the-City: The Supermarket in Germany', in ICREFH IX, 2007, 165–75.

Meldau, R. *Zeichen. Warenzeichen, Marken, Kulturgeschichte und Werbewert graphischer Zeichen* [Trademarks, Trade Names, Cultural History and the Value of Signs in Graphic Advertising], Berlin, 1967.

Müller-Brockmann, J. (ed.), *Das Fotoplakat von den Anfängen bis zur Gegenwart* [The Photo Poster from its Origins to the Present], Aarau, 1989.

Naumann, F. *Kunst im Zeitalter der Maschine* [Art in the Age of Machinery], 2nd ed., Berlin, 1908.

Neuloh, O. and Teuteberg, H. J. *Ernährungsfehlverhalten im Wohlstand. Ergebnisse einer empirisch-soziologischen Untersuchung heutiger Familienhaushalte* [Dietary Excess in Prosperity. Results of an Empirical Sociological Investigation of Today's Family Households], Paderborn, 1979.

Pierenkemper, T. (ed.) 'Die bundesdeutsche Massenkonsumgesellschaft 1950–2000' [The West German Consumption Society, 1950–2000], in *Jahrbuch für Wirtschaftsgeschichte* 2, Berlin, 2007.

Prinz, M. (ed.), *Der lange Weg in den Überfluss. Anfänge und Entwicklung der Konsumgesellschaft in der Vormoderne* [The Long Road to Abundance. Origins and Development of Consumer Society before Modernity], Paderborn, 2003.

Reichwein, S. *Die Litfaßsäule* [The advertising pillar], Berlin, 1980.

Reinhardt, D. *Von der Reklame zum Marketing. Geschichte der Wirtschaftswerbung in Deutschland* [From Advertising to Marketing. An Economic History of Advertising in Germany], Berlin, 1993.

Reulaux, F. *Kunst und Technik*, 1884 [Art and Technology, 1884], reprinted in Richter, K.T. (ed.), *Das Kunstgewerbe die Gewerbe und die Kunstgewerbeschulen und der Marken- und Erfinderschutz* [The Arts and Crafts Industry and the Arts and Crafts Schools, Brands and Protection for Inventors], Wien, 1969.

Riepenhausen, A. *Blechplakate – die Geschichte der emaillierten Werbeschilder* [Sheet Metal Poster: the History of Enamel Advertising ssigns], München, 1979.

Schindler, H. *Monographie des Plakats. Entwicklung, Stil, Design* [Monograph on Poster. Development, Style, Design], München, 1972.

Schlegel-Mathies, K. 'Anfänge der modernen Lebens- und Genussmittelwerbung: Produkte und Konsumgruppen im Spiegel von Zeitschriftenannoncen' [Origins of Modern Food and Beverage Advertising: Products and Consumer Groups in the Mirror of Magazine Advertisements], in Teuteberg, H.J. (ed.), *Durchbruch zum modernen Massenkonsum. Lebensmittelmärkte und Lebensmittelqualität im Städtewachstum des Industriezeitalters* [Breakthrough to Modern Mass Consumption. Food Markets and Food Quality in Urban Growth of the Industrial Age], Münster, 1987, 277–308.

Schmidt, J. 'How to Feed Three Million Inhabitants: Berlin in the First Years after the Second World War, 1945–1948', in ICREFH IX, 2007, 63–73.

Schwarz, J. *Bildannoncen um die Jahrhundertwende* [Image Advertisements around the Turn of the Century], Frankfurt am Main, 1990.

Sombart, W. 'Wirtschaft und Mode. Ein Beitrag zur modernen Bedarfsgestaltung' [Economy and Fashion. A Contribution to Modern Design Requirements], in *Grenzfragen des Nerven- und Seelenlebens* 12, Wiesbaden, 1902.

Spitzer, L. 'American Advertising Explained as Popular *Art*,' in *Essays on English and American Literature*, Princeton, 1962, 248–77.

Teuteberg, H.J. *Die Rolle des Fleischextrakts für die Ernährungswissenschaften und den Aufstieg der Suppenindustrie. Kleine Geschichte der Fleischbrühe* [The Role of Meat Extract in the Nutritional Sciences and the Rise of the Soup Industry. A Short History of Meat Broth], Stuttgart, 1990.

Teuteberg, H.J. 'Strukturmerkmale multimedialer Revolutionierung von Wirtschaft, Gesellschaft und Kultur an der Wende zum 21. Jahrhundert' [Structural Criteria of the Multimedia Revolution in Society and Culture at the Turn of the 21st Century], in Teuteberg, H.J. (ed.), *Vom Flügeltelegraphen zum Internet. Geschichte der modernen Telekommunikation* [From the First Telegraph to the Internet: a History of Modern Telecommunications], Stuttgart, 1998, 294–409.

Wischermann, C. *Advertising in the European City: Historical Perspectives*, Aldershot, 2007.

Chapter 7

Labelling Standard Information and Food Consumption in Historical Perspective: An Overview of State Regulation in Spain 1931–1975

Gloria Sanz Lafuente

Introduction

Labelling represents the most important communication instrument between food producers and consumers and a mechanism to protect consumers against deception and fraud. Historically, food-labelling information has included three levels of communication with consumers, which were moulded by different historical backgrounds. The first development was related to the commercial framework and market segmentation in a growing mass market: image, product distinctiveness and brand development.[1] The second appeared later with information for public safety control: the name and address of the producer and retailer, net quantity of contents, name of the food, and statement of ingredients. A third was recently developed in relation to food quality and nutrition and included nutrient reference values of the products. Foods have to be labelled with an indication of their nutritive value, including amounts of fat, protein, carbohydrates or alcohol.[2] For example, a decree on indications of nutritive values was issued in the Federal Republic of Germany in 1977.

Marketing and information related to food safety and quality showed an important feedback relationship, especially in the 1960s and 1970s when the golden age of advertising started and the notion of consumer information turned to protection in the consumer policies of some countries.[3] In 1978, K. Fraser, from Unilever's marketing division, recognized that an 'economic function of food labelling' was 'its role in marketing and commercial promotion'. W. Roberts of the International Association of Consumers' Unions in the United Kingdom talked at the same

---

1 Jones and Morgan 1994.

2 Schulze and Mücke 1980: 397. For contemporary debate, see the contribution of Gun Roos in this volume.

3 Hilton 2006: 45–61, Theien 2006: 29–44, Kleinschmidt 2006: 13–28, Trumbull 2001: 261–82, Trumbull 2006.

meeting about 'the social function of food labelling: consumer information'.[4] On the one hand, marketing manipulated information on the label to increase its sales of food; on the other hand, new rules generated new marketing strategies in the food industries.[5] Retailers and manufacturers recognized, for example, the growth in consumer demand for healthy food and responded to it in the 1980s.

There were significant differences between countries with regard to foodstuff labelling. In 1906 the Federal Food and Drugs Act in the USA stated only 'that any declaration on the label of a food had to be truthful and not misleading. Significantly, statements or claims about a food made anywhere but on the label were not subject to the act'.[6] The subsequent Federal Food, Drug and Cosmetic Act was passed in 1938. This law prohibited false or misleading statements in food labelling and required affirmative labelling of food which included the information specified in the statute (the name and address of the manufacturer, net quantity of contents, name of the food, and statement of ingredients), but did not find a great need to describe food content to consumers.[7] In Germany, the *Lebensmittelgesetz* of 1927 had already established important compulsory food labelling that was extended in 1935 by the Food Labelling Decree, while food-labelling regulations were introduced in Switzerland in 1936.[8]

The establishment of voluntary and compulsory food labelling is an historical process by various social and political participants including consideration of the current scientific knowledge about nutrition and food, political decisions, consumption cycles and the demands of consumers and producers. The aim of this chapter is to present an overview of the development of compulsory labelling standards of processed food in Spain since 1931 up to the end of the Dictatorship in 1975, as well as of its relationship to the different political and economic periods.

Spanish Food Labelling in Consumption Cycles and Political Context, 1931–1975

The identification of a product on the market by means of a distinctive brand label was a feature of the definition of 'industrial property' long before it

---

4   Fraser 1980: 31–3, Roberts 1980: 23–30.

5   For food quality and the legal position in the United Kingdom see Oddy, ICREFH IX: 91–103, also French and Phillips 2000. In relation to the nutritional labelling discussion in the United Kingdom, Freckleton has suggested that 'The extent to which a manufacturer provides nutritional labelling depends largely upon the types of food being produced'. Freckleton 1985: 8–9.

6   Summers 1999: 3.

7   Institute of Medicine 2003: 20.

8   'Im Rahmen dieses Gesetzes wurde erstmals die Pflicht zur Kennzeichnung von Lebensmitteln postuliert', [Under this law the labelling of foods was first postulated], Grube 1997: 9, Schulze and Mücke 1980: 395, Haesler 1980: 449.

identified information, consumer protection or food safety. Industrial property rights, and within them the right to a brand associated with a trademark, took on early relevance in the formation of an ever-wider and competitive industrial food market.[9] The creation of distinctive brands with labels by the late nineteenth century represented a mechanism for communication with consumers, market segmentation and the regulation of competition among producers, thus becoming the basis of inter-business conflicts. A distinctive identification was a primary marketing tool for releasing a new product on the market.[10] The use of external labelling up to the 1930s within the area of industrial property had been legally regulated with the purpose of guaranteeing ownership and combating fraud. Thus, instead of having a single standard of labelling, there were multiple stipulations affecting certain specific food products. However, the legal information supplied by a label in this context was voluntary and limited by narrow legal bounds. Defining a brand did not mean indicating to consumers the quality differences of products, despite the use of corporate standard references. The additional information that was supplied featuring the properties of the product, was left to the judgment of the food and farming industry. It became a tool for marketing but not for informing the public. Moreover, the consumer could not participate in the construction of this information within the legislation on industrial property. Together with the differentiation of industrial property and food marketing, another function of food labelling appeared that was related to the first advances in public health and nutrition.[11] Although misleading publicity and usurpation of brands were prosecuted, the time was still far off when obligatory legislation of labelling for all food products. Up to that time the provisions stipulated that manufacturers could be subject to spot inspections by local officials; misleading publicity and fraud was supposedly combated, although in the city of Zaragoza, in 1921 only 3 per cent of prosecutions were for this offence.[12] There were also regulations on the labelling of some specific products which were partial and even local in nature. Provisions referring to a compulsory sanitary labelling were still in their infancy and there was not a standard for all food products.

During the short-lived democratic framework of the Second Republic (1931–1936), the tendency towards consumption growth was maintained and slightly diversified, with foodstuffs consumption showing a slight downward trend in percentage terms. During this period, the general bases of earlier food labelling legislation were maintained, showing the lack of a single and well-defined standard for all products, though there were some significant improvements. A greater amount of financing for the health area was the first of these signs, and the First National Health Congress of 1934 indicated a growing academic interest in

---

9   Saiz 1999: 265–302.
10   Macleod 1988, Vaughan 1956, Penrose 1962.
11   About food control, see Sanz Lafuente 2006.
12   Sanz Lafuente 2006.

foodstuff control.[13] Some specific regulations, such as those referring to wine, or the creation of the Institute of Food Hygiene were focused on increasing producer transparency towards consumers.[14] From the legislative point of view, there was definite progress with the intention to 'subject all food preparations circulating in the market to checks and verifications carried out by said institute', with the purpose of preventing 'the abuse of consumer good faith in the sense of acceptance of the validity of all the publicity claims in regard to the quality of the preparation'.[15] Among the new measures was the need for all products to have prior authorization of the General Inspection Office of the Pharmaceutical and Chemical Industries before their introduction into the market. This authorization was only to be given after a report from the Institute, and the registration number of the authorization had to be displayed on the packaging. This law however was never put into practice amidst the chaos of the civil war, nor during the Dictatorship that followed.[16]

The strict measures for chocolate were maintained.[17] In other cases, the measures were subject to diverging economic and political interests. In 1933, the Ministry of Finance authorized the leaseholders of the Cádiz Public Warehouse 'to export and bottle national olive oil under foreign brand names' because this enhanced oil exports, thereby encouraging exportation of a product that had been subject to restraints during the crisis. In 1935, the Industry and Trade Ministry decreed the prohibition of 'exporting olive oil under brand packaging that omitted or did not explicitly and visibly state the Spanish origin of the merchandise' so as to avoid practices which might affect the 'good name of Spanish oil'.[18] A basic change in favour of greater regulation of labelling occurred with the 1932 Wine Statute. In the first place, no product or mixture for use as wine could be manufactured, promoted or commercially distributed, if the packaging did not specify the quantitative composition of the product, nor could a place name be given to a wine under the pretext that it was of a similar kind if that wine had not been grown, prepared and manufactured in the respective area. Also, on sale, the package had to specify with visible lettering its alcoholic content by percentage and its price per litre. Bottled wines had to carry a label with the name and registration number of the bottling firm and the town where the bottling cellar or facility was located.[19]

After the Civil War (1936–1939), food consumption shrank amidst problems of overpricing and hunger resulting from Franco's economic policy of autarky.

---

13   For research about the lack of vitamins, food hygiene and the chemical composition of food, see Primer Congreso Nacional de Sanidad 1934.

14   Decree of 10 April 1937, Order of 21 October 1937, Order of 15 February 1938.

15   *Gaceta de la República* n° 47 16-02-1937.

16   Barciela 2003.

17   'Orden sobre composición y fabricación del chocolate' [Order on the Composition and Manufacture of Chocolate], *Gaceta de Madrid* n° 353, 19-12-1935.

18   Orden de 6 de Mayo de 1933, *Gaceta de Madrid* n° 139, 19-05-1933, Decreto de 26 de Julio de 1935, *Gaceta de Madrid* n° 212, 31-07-1935.

19   González 2005: 80–82.

If consumption level per inhabitant in 1935 is taken as a reference, this was not regained until 1957 as Figure 7.1 shows. Consumption was merely the expression of other variables such as the fall in the Gross Domestic Product per head. Together with this drop, the diversification of consumption structure worsened compared with the 1930s and most of the family budget was centred on food and beverages. In 1935, 56 per cent of the private domestic consumption in Spain consisted of food, beverages and tobacco. After the fluctuations caused by the Civil War, this percentage increased, reaching 64 per cent in 1950. The 1935 percentage share – one year before the start of the Civil War – was not reached again until 1955. In short, in Spain up to the mid-1950s, there was a reduction of both consumption levels and composition of consumption, with a disproportionate share of expenditure earmarked for the purchase of foodstuffs. The rationing cards introduced in 1939 were not to disappear until 1951, and attempts at market and foodstuffs production control led to the emergence of a busy regulation-free and health-control-free black market. The Sponsorship Council for the Study of Food and Nutrition Hygiene created in 1947, showed in its general outlook and in the measures it took a markedly different spirit from the 1937 regulations. It exhibited a more study-oriented than interventionist policy.

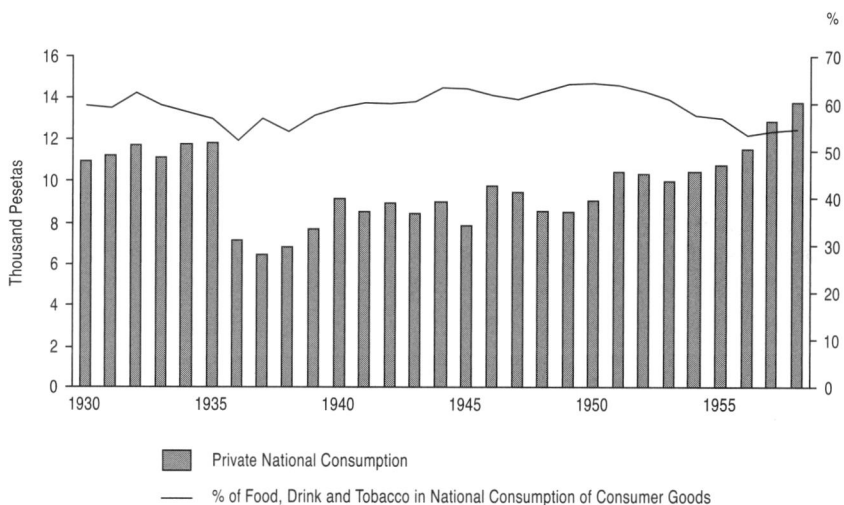

Private National Consumption

% of Food, Drink and Tobacco in National Consumption of Consumer Goods

Figure 7.1 Real Private National Consumption (per person) and Percentage of Food, Drink and Tobacco in the Private Consumption of Goods: Spain 1930–1958 (1958 pesetas and per cent)

Sources: Maluquer de Motes 2005, 1279, 1285–6.

Within this deficit-prone context, the quality of demand became a secondary consideration, as was evidenced in the Law for the Foundation of National Health in November 1944. Among the amendments to the Dictatorship's original project were those related to the General Health Board and the General Commissariat for Provisioning and Transport at a time of rationing and food shortages. One of the amendments stated that 'the equitable distribution of available stocks among all Spaniards...cannot possibly be conditioned by what the General Health Board qualifies as food, bearing in mind only sanitary considerations, since in many cases due to lack of stocks, technology must adapt and subordinate itself to prevailing realities'. As this document indicated, had the General Health Board set the sanitary conditions for foodstuffs and had the ration mix been subordinated to these conditions, it would have 'produced a natural problem of provisioning'. The amendment was included within the law, and attributed to the General Health Board but 'without affecting the faculties attributed to the General Commissariat for Provisioning'.[20] In the context of scarce, expensive and controlled food products, quality considerations would be subordinated to obtaining the necessary amounts of food. Legislation in this case was designed to control the market rather than to make it safe for the consumer.

Legislative activity requiring consumer information on foodstuffs reappeared in the mid-1950s, triggered by increased consumption growth. Foreign capital inflows were directed at food-producing agriculture and the Instituto Nacional de Industria started to intervene in the foodstuffs industry.[21] The incipient opening up of the local economy demanded changes in order to drive trading activity abroad in the shaky primary agricultural foodstuffs industry, bearing in mind, moreover, that foreign food companies were establishing themselves in the Spanish market. The country's entry into the Food and Agricultural Organization in 1950 – the first international organization to recognize the Dictatorship, even before the United Nations or the International Monetary Fund – also signalled a new foreign relations framework. The Inter-Ministerial Commission created in 1955 for the technical health regulation of all the industries included within the Syndicate of the Foodstuffs and Beverages Industry was initially formed by representatives of the Ministry of Industry and Commerce (The General Commissariat for Provisioning and Transport) and the Ministry of Government (General Health Board), but then incorporated representatives from the Ministries of Agriculture, Finance and Labour. Only one of the ten members was an official for the health area. Within the objectives of the Law was included that of serving 'the most basic standards of health', and it originally stressed the need to 'avoid clandestine manufacture' of foodstuffs possessing no brand and subject to no controls, which were sold in bulk, because this would negatively affect industrial interests.[22]

---

20 Archivo del Congreso de Diputados, *Serie General de Expedientes. Proyecto de Ley de Bases de Sanidad Nacional 1944, 955–1.*

21 Barciela, López and Melgarejo 2004: 127–62.

22 Sindicato Nacional de la Alimentación 1963.

With regard to foodstuffs labelling, technical health-regulatory activity was associated with specific products. This activity included a number of different products over time, but significantly left out the mass-consumption items (cooking oil, wine or sugar, for example). Neither did it incorporate other products whose consumption had recently expanded, such as preserves. In its legal configuration also different criteria were applied with regard to information to be supplied to the consumer and there were different requirements for different products. The wide disparity of criteria and heterogeneity created 'visible' and 'invisible' information, depending on the product concerned (Table 7.1). For example, information on components or the use of preservatives and artificial sweeteners differed in the regulations according to the products concerned. All the foregoing was framed within the dictatorial legislative context, restricting public debate, quality being driven by consumer education and industrial interests.

Table 7.1   Food Labelling Requirements in the Technical Health
            Regulations of Some Products, 1956–1963

Biscuits *26/02/1962	Juice; soft drinks *13/13/63	Flavouring *05/04/1963
Registration in public sanitary register	Yes; yes	Yes
Name of manufacturer	Yes; yes	Yes
Registered trade mark	Yes; yes	Yes
Address	Yes; yes	Yes
Registered composition	Technical name and quality of the product	Name of the product
Weight of the packaged biscuits	Net weight; yes	–
Number in public sanitary register	Yes; no	Yes
Number of the manufacturer	Yes; yes	–
	Proportion and kind of sweetener, preservative; authorized sweetener	'For food use'

Source: Technical health regulations. * Data from the BOE, 1956–1963

The 1959 Stabilization and Liberalization Plan signalled a turnaround in the economic policies of the Dictatorship in Spain. These were the days when 'developmentalism' appeared on the scene and economic variables began to recover, showing growth after the hard years of autarchy. It is certain that the years of hunger and rationing were still present in the short-term memory of the 'growth society' during the 1960s and 1970s. During this complex period, economic growth accelerated, combined with a progressive increase in levels of consumption.[23] Despite the wage growth figures observed as from the late-1950s, most of the new personal income was invested in basic consumer goods such as food. In 1958, 55 percent of private consumption expenditure per head per year corresponded to food and beverages. By 1974, this had dropped to 38 per cent which remained a high figure, as can be seen in Figure 7.2. Together with this transformation, internal migratory flows generated an urban market that was both further away from farm producers and increasingly segmented, in which a greater quantity of artificially-prepared foodstuffs was consumed.[24] The first local standard for food labelling was legislated in the Spanish Foodstuffs Code of 1967. Only in 1975 did a general labelling and publicity standard for packaged products become law. Food information in the 1960s was related more to food marketing than to information related to food safety and quality since it was uncontrolled.

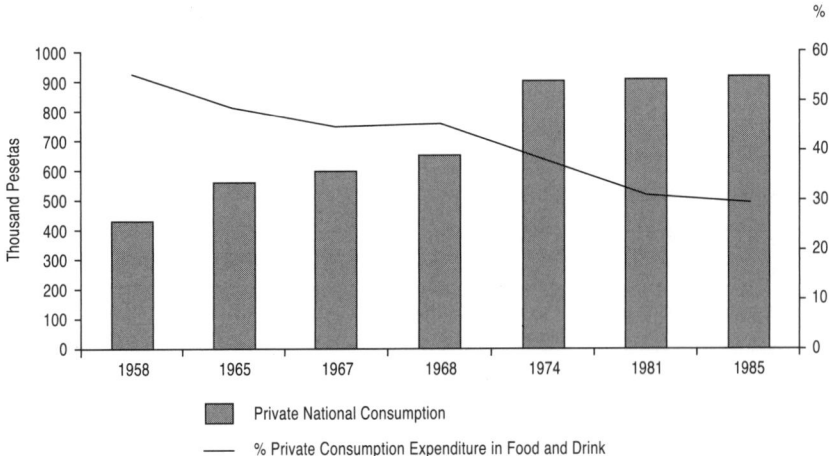

Private National Consumption

—— % Private Consumption Expenditure in Food and Drink

Figure 7.2   Total Private National Consumption (1999 pesetas per person) and Percentage of Expenditure on Food and Drink in Private Consumption (per person): Spain 1958–1985

Source: Maluquer de Motes, 2005.

In 1950 there was a meeting of the Joint FAO/WHO Expert Committee on Nutrition. This committee recognized that:

23   Cussó and Garrabou 2007: 69–100.
24   Andrés 1977: 149–58, Fundación FOESSA 1972: 225–34, Fundación FOESSA 1976: 508–28.

Food regulations in different countries are often conflicting and contradictory. Legislation governing preservation, nomenclature and acceptable food standards often varies widely from country to country. New legislation not based on scientific knowledge is often introduced, and little account may be taken of nutritional principles in formulating regulations.[25]

The Codex Alimentarius, or the food code, was an important instrument that emerged after the Second World War. It combined science, consumer health protection and the interests of the growing international food trade. The Codex represented an attempt at establishing a uniform international food standard together with guidelines and recommendations. Spain was one of the 30 countries that participated in the first session of the joint FAO/WHO Codex Alimentarius Commission held in Rome in 1963.[26] In 1969 the Codex Committee on Food Labelling published general international standards for the labelling of pre-packaged foodstuffs.[27] These guidelines take into account work done by the Codex Committee on Food Labelling in the Codex Alimentarius Commission. These were only proposals. They suggested that there should be no limit on the amount of information which might be given, but there was no prescribed format. Despite these limitations, the guidelines were an improvement because they proposed a standard that included a list of ingredients and components.

With regard to labelling in Common Market countries, Leon Klein, the head of the consumer division of the European Common Market, pointed out in 1972: 'the work of harmonizing legislation is an essentially pragmatic task, and there is a need to create a clear and easily understood line of procedure, product by product'.[28] In 1978 a new Directive was adopted.[29] The EEC proposal introduced required legends which were not listed in the Codex standard, such as the 'use by date' and any special storage conditions or conditions of use.[30] The EEC Directive was more accurate than the Codex but like the Codex, prohibited misleading statements as well as health claims, and prohibited 'trade in those products which do not comply with the provisions of this Directive four years after its notifications' (Article 22).

The Spanish Foodstuffs Code (SFC) was created in the context of the efforts of a number of countries to introduce their own foodstuffs codes, and because of the regime's wish to show conformity with international standards. Spain's participation in the First Session on the Codex Alimentarius in Rome and the simple ambition

---

25   FAO 1999.

26   Comisión Conjunta FAO/OMS Codex Alimentarius 1963: 12–3.

27   Comisión del Codex Alimentarius 1969.

28   Primera Asamblea Nacional de Consumidores 1972: 73.

29   *Official Journal of the European Communities* N° L33, 8th February, 1979. Council Directive of 19 December 1978 on the Approximation of the Laws of the Member States Relating to the Labelling, Presentation and Advertising of Foodstuffs for Sale to the Ultimate Consumer.

30   Delville 1980: 14–151.

of Franco to join the European Common Market, led to the incorporation in the legislation of international standards that were considered a fundamental condition for foreign acceptance. International trade and the adaptations necessary to allow the greater penetration of the Spanish market by foreign foodstuffs also made conformity with international standards necessary.[31] The international activity of institutions such as the FAO and the WHO was underlined by the creation of the subcommittee of experts whose task it was to formulate the SFC in 1960.[32] This group of experts included both the National Farming Industry Commission and a number of professional nutritionists specializing in health and dietetics in the area of veterinary and human medicine. Its work was difficult, and was carried out in the context of a Dictatorship and conflicting industrial interests, as when the Association for Vegetable Industry Research criticized the code for its lack of uniformity and as a 'limitation for technical progress'.[33]

The publication of the Code in 1967 amounted to the incorporation of the guidelines that had been proposed by the WHO and the FAO after the Second World War. Spain joined the international attempt to regulate systematically the conditions for the production and distribution of foodstuffs. However, there were problems characteristic of Dictatorial 'developmentalism'; the main one was a delay in enforcement until 1974 and the non-implementation of a new set of technical health regulations. Thus, it became merely a series of general precepts which remained as the basis for future specific regulatory efforts, and which were therefore not applicable. The general labelling and publicity standard for packaged products established a distinction between generic and obligatory labelling (brand name, commercial name, legal domicile, denomination, net contents, country of origin and sanitary registration number), academically correct ('facultative') labelling (neither wrong nor misleading) and another type, called specific labelling. In this last, which was optional in nature, the list of ingredients and additives appeared as well as the date of packaging and sell-by date, the lot number and the commercial category 'for specific foodstuffs that require this information'. No mention was made of which specific products were to be included.[34] For purposes of comparison, in the case of Italy, albeit with the exception of some products, in 1962 a list of the ingredients in decreasing order of percentage composition was required.[35]

If one takes for reference purposes the general recommendations of the FAO/WHO committee, those of the SFC and some rulings that affected specific products, and that were put into practice in those years, some significant differences of judgement become apparent. Spain did not require a list of ingredients, though this

---

31   Kermode 1980: 237–57.

32   Sindicato Nacional de la Alimentación 1963: 11.

33   *Alimentaria. Revista de Tecnología e Higiene de los Alimentos*, n° 19, Mayo-Junio, 1968: 3.

34   Decreto 336/1975, BOE n° 60 11 de Marzo de 1975.

35   Mor 1980: 407.

was necessary for some products subject to specific technical health regulations, but its absence from the general recommendations is noteworthy.

During the 1960s, the years of growth made increased food intake, both in energy value and nutrient content, synonymous with development and a good foodstuffs policy.[36] However, proper dietary-type labelling was not considered at the political level during these years, nor did it appear in the guidelines of the FAO's Codex Alimentarius in 1969. Moreover, in the early-1960s, new advances in the food processing sector were generating the demand for consumer protection against non-visible contaminants or pollutants and for information on product composition in the academic media and international organizations. Antibiotics used for preserving food were an item of interest for producers at the VI International Symposium on foreign substances in foodstuffs, which was held in Madrid in 1962, organized by the Commission Internationale des Industries Agricoles and by the Bureau Internationale Permanent de Chimie Analytique. However, the chemical composition of antibiotics was a matter of concern for the health science community and international organizations were demanding proper regulation.[37] Control of pesticide contamination began to form part of recommendations in the WHO's technical reports in 1962.[38] Something similar was happening in regard to foodstuff additives and the assessment of their carcinogenic potential or the improper use of nitrates and nitrites for the preservation of meat products and to enhance their colour or prevent bacterial growth.[39] The road from scientific debate towards official regulation was a rocky one and the labelling information provided in Spain did not achieve homogeneity either. In the technical health regulations of the 1950s, which came into force in the 1960s and 1970s, information was required on preservatives and sweeteners used for soft drinks, but not for biscuits or preserves. Luis González Vaque, the Vice President of the Spanish section of the European Association for Food Health Rights (AEDA) and a legal adviser to the Spanish Agency for Promoting the Exportation of Foodstuffs Products (PROEXPA), made the following comment at the international meeting on labelling and publicity that was held in Brussels in 1978:

> I must admit that we have not set in motion a general type of regulation that requires the listing of additives on foodstuff product labels. I would like to briefly point out that we have passed a Foodstuffs Code as a major joint project destined for application to all foodstuff products. This will be a basic legal text to be completed through different sets of technical health regulations...of a vertical (obligatory) nature...Only specific provisions (of the regulations) stipulate a listing of additives on the label. [40]

---

36   FAO 1969.
37   FAO/WHO 1970, Real Academia de Medicina de Valencia 1968.
38   Ministerio de Agricultura 1961.
39   WHO 1962, FAO/WHO 1961, FAO/WHO 1974, Sainz 1971.
40   González Vaque 1980: 113–4, González Vaque 1980: 363–9.

Within this context and in the early 1970s, the first steps were taken towards the creation of Spain as a consumer society, and after an inaugural meeting in Barcelona, the first National Assembly of Consumers was held in 1972.[41] This was attended by 37 consumer associations, as well as other groups such as family associations or consumer cooperatives, and was presided over by Enrique Villoria. In the assembly, one of the participants, José María del Rey Villaverde, underlined that although there had been a quantitative increase of consumption, 'the current demand for quality', meaning the ability and capacity to 'demand as Europeans and not as citizens of an underdeveloped society' was not keeping pace with the consumption boom. Underlying this fact he noted the lack of anti-fraud and dietary education. Likewise, full implementation of the Spanish Food Code was demanded, while some food-industry representatives demanded information-type labelling, and consumer representation in government agencies, as was already the case in several European countries.[42] However, official consumer recognition was still a long way off. The General Law for the Defence of Consumers and Users was only passed in 1984. Implementation of the Spanish Food Code received support in 1981, during the complex transition years. After the major oilseed rape (colza) mass poisoning scare affecting thousands, a full session of the Spanish Congress passed a non-legally enforceable resolution to enhance consumer protection against foodstuffs frauds. In 1982, the General Standard for packaged product labelling, presentation and publicity was passed by Congress. This programme of activity was to culminate with the SFC's development in 1985, coinciding with the country's entry in the EEC.

Conclusions

From 1931 to 1975, extremely diverse political and economic cycles succeeding one another limited the development of labelling standard information. After the civil war, food information for consumers was reduced in a context of scarcity with food production controlled by the Dictatorship and leading to a growing black market for food.[43] The multiplicity of regulations and the lack of a food-labelling standard continued until product-based regulations appeared in the mid-1950s. The development of a food labelling standard was just one of the items involved in the evolution of a consumer market for processed food products, with the expansion of the urban market and the emergence of class and variety-based segmentation in the competitive market. Information about food standards increased in the 1960s in the context of a growth of processed-food consumption, international trade and competitive production by manufacturers. Food-labelling standards could only be summarized as a single body of recommended practice with the drawing up of the

---

41   Primera Asamblea Nacional de Consumidores 1972, OCU 1976.
42   Asamblea Nacional de Consumidores 1972: 23–28, 34, OCU 1976.
43   Barciela 2003.

Spanish Food Code in 1967 and with the law passed in 1975. In the increasing food market of the 1960s and 1970s some manufacturers stimulated the search of the economic asset of acquired prestige with the marketing tools: labelling became a part of their new tools. Professional scientists and the first consumer movement in the 1970s rejected the slow enforcement of the SFC. Slow enforcement was the basic feature in the application of general standards to foodstuffs labelling during the Dictatorship.

References

Andrés, F. *Las Bases Sociales del Consumo y del Ahorro en España* [The Social Bases of Consumption and Saving in Spain], Madrid, 1977.

Asamblea Nacional de Consumidores *Primera Asamblea Nacional de Consumidores* [First Consumers' National Assembly], Madrid, 1972.

Barciela, C. *Autarquía y Mercado Negro: el Fracaso Económico del Primer Franquismo, 1939–1959* [Black Market Autarky and the Economic Failure of Franco], Barcelona, 2003.

Barciela, C., López, I. and Melgarejo, J. 'La Intervención del Estado en la Industria Alimentaria durante el Franquismo (1939–1975)' [State Intervention in the Food Industry during the Franco Period], *Revista de Historia Industrial* 25, 2004, 127–62.

Calleja, J.L. *La Publicidad y la Marca, Armas Contra el Fraude* [Advertising and Branding, Weapons against Fraud], Madrid, 1972.

Comisión Conjunta FAO/OMS Codex Alimentarius, *Informe del Primer Periodo de Sesiones, Roma, 25 de Junio–3 de Julio 1963* [FAO/WHO, Report of the First Session] Rome, 1963.

Comisión del Codex Alimentarius, *Norma General Internacional Recomendada para el Etiquetado de los Alimentos Preenvasados, Programa Conjunto FAO/OMS* [Recommended International Standards for the Labelling of Prepackaged Foods], Rome, 1969.

Cussó, X. and Garrabou, R. 'La Transición Nutricional en la España Contemporánea: las Variaciones en el Consumo de Pan, Patatas y Legumbres (1850–2000)' [The Nutrition Transition in Contemporary Spain: Variations in the Consumption of Bread, Potatoes and Vegetables], *Investigaciones de Historia Económica* 7, 2007, 69–100.

Delville, M.R. 'Comparative Outline of International Regulations Relating to Food Labelling', in Institut d'Études Européennes (ed.) *Labelling and Advertising Regulations on Food Products*, Brussels, 13 and 14 April, 1978, Brussels, 1980, 147–51.

FAO, *Políticas de Alimentos y Nutrición* [Food and Nutrition Policy], Rome, 1969.

FAO/WHO, *Evaluación de los Peligros de Carcinogénesis que Entrañan los Aditivos Alimentarios* [Assessing the Carcinogenic Hazards of Food Additives], Geneva, 1961.

FAO/WHO, *Normas de Identidad y Pureza para los Aditivos Alimentarios y Evaluación de su Toxicidad: Diversos Antibióticos* [Identity and Purity Standards for Food Additives and Evaluation of Toxicity: Various Antibiotics], Geneva, 1970.

FAO/WHO, *Evaluación de Ciertos Aditivos Alimentarios* [Evaluation of Certain Food Additives], Geneva, 1974.

FAO/WHO *Understanding the Codex Alimentarius,* 1999. Available at: http://www.fao.org/docrep/W9114E/W9114e00.htm [accessed 22 February 2009].

Fraser, K. 'The Economic Function of Food Labelling, Marketing and Commercial Promotion' in Institut d`Études Européennes (eds), 31–8.

Freckleton, A. *Who is Shaping the Nutritional Label? A Report on the Activities of Government, Manufacturers, Retailers and Consumers,* Bradford, 1985.

French, M. and Phillips, J. *Cheated not Poisoned? Food Regulations in the United Kingdom, 1875–1938,* Manchester, 2000.

Fundación FOESSA, *Estudios Sociológicos sobre la Situación Social de España, 1975* [Sociological Studies of the Social Situation in Spain], Madrid, 1976.

Fundación FOESSA, *Síntesis del Informe Sociológico sobre la Situación Social del España 1970* [Sociological Synthesis of the Social Situation of Spain], Madrid, 1972.

González Vaque, L. 'La Réglamentation de l'Etiquetage et de la Publicité des Produits Alimentaires en Espagne' [Spanish Regulations on Labelling and Advertising of Food Products], in Institut d`Études Européennes (eds), 363–9.

González Vaque, L. 'L'Indication des Additifs sur l'Etiquette' [The Disclosure of Additives on the Label], in Institut d`Études Européennes (eds), 113–4.

González, F. *El Régimen Jurídico del Etiquetado de Vinos* [The Legal System of Wine Labelling], Barcelona, 2005.

Grube, C. Verbraucherschutz durch Lebensmittelkennzeichnung? Eine Analyse des deutschen und europäischen Lebensmittelkennzeichnungsrechts [Consumer Food Labelling? An Analysis of German and European Food Labelling laws], Berlin, 1997.

Haesler, M. 'La Réglementation de l'Etiquetage et de la Publicité des Produits Alimentaires en Suisse' [The Regulation of Labelling and Advertising of Food Products in Switzerland], in Institut d`Études Européennes (eds), 449–56.

Hilton, M. 'Consumer Protection in the United Kingdom', *Jahrbuch für Wirtschaftsgeschichte* 1, 2006, 45–61.

Institut d'Études Européennes (eds) *Labelling and Advertising Regulations on Food products, Proceedings of a Colloquium, Brussels, 13 and 14 April, 1978,* Brussels, 1980.

Jones, G. and Morgan, N.J. (eds) *Adding Value: Brands and Marketing in Food and Drink,* London, 1994.

Kermode, C.O., 'Provisions for Labelling and Advertising of Foodstuff adopted by the FAO/WHO Codex Alimentarius Commission', in Institut d'Études Européennes (eds), 237–57.

Kleinschmidt, C. 'Konsumgesellschaft, Verbraucherschutz und Soziale Marktwirtschaft. Verbraucherpolitische Aspekte des "Modell Deutschland" (1947–1975)' [Consumer Society, Consumer Protection and the Social Market Economy: Political Aspects of the German Consumer Model], *Jahrbuch für Wirtschaftsgeschichte* 1, 2006, 13–28.

Maluquer de Motes, J. 'Consumo y Precios' [Consumption and Prices], in Carreras, A. and Tafunell, X. (eds) *Estadísticas Históricas de España* [Historical Statistics of Spain], Bilbao, 2005, 1248–96.

Martín Serrano, M. *Publicidad y Sociedad de Consumo en España* [Advertising and Consumer Society in Spain], Madrid, 1970.

Ministerio de Agricultura, *VIᵉ Symposium sur les Substances Etrangères dans les Aliments, Madrid, 10–15 Octobre 1960* [Sixth Symposium on Foreign Substances in Food], Madrid, 1961.

Mor, B. 'La Réglementation de l'Etiquetage et de la Publicité des Produits Alimentaires en Italie' [Regulation of Labelling and Advertising of Food Products in Italy], in Institut d'Etudes Européennes (eds), 407–13.

OCU, *Estatutos de la Organización de Consumidores y Usuarios Aprobados por la Asamblea Fundacional el 30 de Julio de 1975* [Regulations of the Organization of Consumers and Users Approved by the Assembly], Madrid, 1976.

Oddy, D.J. 'Food Quality in London and the Rise of the Public Analyst, 1870–1939' in ICREFH IX, 94–103.

Penrose, E.T. *Teoría del crecimiento de la empresa* [The Theory of Enterprise Growth], Madrid, 1962.

Primer Congreso Nacional de Sanidad *Extractos de las Comunicaciones* [Excerpts of Communications], Madrid, 1934.

Real Academia de Medicina de Valencia, *Utilización de Antibióticos en la Conservación de Alimentos* [Use of Antibiotics in Food Preservation], Valencia, 1968.

Roberts, W. 'The Social Function of Food Labelling: Consumer Information and its Extent', in Institut d'Etudes Européennes (eds), 23–30.

Sainz Cidoncha, F. 'Problemática de la Utilización de Nitratos y Nitritos en la Tecnología de Productos Cárnicos' [Problems with the Use of Nitrates and Nitrites in Meat Technology], *Alimentaria: Revista de de Tecnología e Higiene de los Alimentos* 8, 40, 1971, 3–20.

Saiz, P. 'Patentes, cambio técnico e industrialización en la España del siglo XIX' [Patents, Technical Change and Industrialization in Nineteenth-Century Spain], *Revista de Historia Económica* 17, 1999, 265–302.

Sanz Lafuente, G. 'Perspectivas de Historia de Seguridad Alimentaria. Entre la Ley y la Práctica Social de la Inspección 1855–1923' [Perspectives on the History of Food Safety: between the Law and Social Practice of Inspection], *Revista de Estudios Agrosociales y Pesqueros* 212, 2006, 81–118.

Schulze, H. and Mücke, W. 'Regulations of Food Labelling and Advertising in the Federal Republic of Germany', in Institut d'Etudes Européennes (eds), 395–400.

Sindicato Nacional de la Alimentación *Comisión Interministerial para la Reglamentación Técnico-Sanitaria de las Industrias de Alimentación* [Interministerial Commission for the Technical Regulation of the Health-Food Industries], Madrid, 1963.

Sindicato Nacional de la Alimentación, *Sub-Comisión de Expertos para el Código Alimentario* [Sub-Committee of Experts for the Food Code], Madrid, 1963.

Summers, J.L. 'Overview of the History of Food Labelling', in *Food Labelling Compliance Review*, IA, 1999, 1–12.

Theien, I. 'From Information to Protection: Consumer Politics in Norway and Sweden in the 1960s and 1970s', *Jahrbuch für Wirtschaftsgeschichte* 1, 2006, 29–44.

Trumbull, G. 'Strategies of Consumer Group Mobilization: France and Germany in the 1970s', in Daunton, M. and Hilton, M. (eds) *The Politics of Consumption: Material Culture and Citizenship in Europe and America*, Oxford, 2001, 261–82.

Trumbull, G. *Consumer Capitalism: Politics, Product, Markets, and Firm Strategy in France and Germany*, Ithaca, 2006.

WHO, *Principios Fundamentales para la Seguridad del Consumidor contra los Residuos de Plaguicidas* [Key Principles for Consumer Safety Against Pesticide Residues], Geneva, 1962.

Chapter 8

Food Labelling for Health in the Light of Norwegian Nutrition Policy

Gun Roos

Introduction

Concerns related to nutrition have shifted from food scarcity to abundance in twentieth-century Europe. In the late nineteenth century undernutrition and food falsification were the main food-related concerns.[1] Today undernutrition is mainly a problem in developing countries, whereas being overweight is viewed as a major public health problem worldwide.[2] The growing incidence of obesity and excess body fat in Europe has put more focus on health and nutrition policy and on finding ways to promote better diets and empower consumers. As a result, nutritional information and health-related food labelling are often included in policy strategies for the promotion of better food choices and diets.

The purpose of this chapter is to describe nutrition policy, nutritional labelling and consumer roles in twentieth-century Norway. Food scarcity and undernutrition characterized the country up to the end of the Second World War, and assuring food security and healthy nutrition for all were still on the political agenda well into the postwar period. Excess bodyweight was first debated in Norway in the 1950s and in recent years it has come to be conceptualized as an epidemic disease. The resulting policies are rather different from those developed for dealing with undernutrition, where the emphasis was upon collective action framed through public responsibility; being overweight and malnutrition tend instead to be dealt with as a matter for the individual. There are different understandings of responsibilities for nutrition and health in ways that were not possible before. In this context, nutritional labelling has received more interest from producers, retailers, health policymakers and consumer organizations. The focus in this chapter will be on simplified nutritional labelling, because there is an ongoing debate about this in Norway.

---

1   See Scholliers, ICREFH IX, 2007, Oddy, ICREFH IX, 2007.
2   WHO 1997.

Food Shortage in Norway

Food shortage was common in Norway from the nineteenth century until the 1950s and this had consequences in terms of public policies. From a public health perspective, the nutritional status of the population can be divided into four main periods: 1860–1920, a phase of scarcity and food poverty; 1920–1940, the vitamin age when the emphasis shifted to micro-nutrients; 1945–1985, an era of abundance but also of malnutrition; and from 1985 to the present, a time of affluence and overnutrition.[3]

The period of scarcity was characterized by poverty, tuberculosis, undernutrition and high infant mortality rates. Although the industrialization of the Norwegian economy had begun, the labour force was diminished by large-scale emigration. The country was dependent on grain imports and, because of this, rationing had to be introduced during the shortages of 1917. The next period was highlighted by the discovery of vitamins. There was also a focus on hygiene resulting in better nutrition and lower infant mortality rates, and the authorities sponsored information campaigns, magazines and exhibitions on food. For example, milk at this time was viewed as a healthy part of the diet, and in 1933 a milk exhibition was arranged to strengthen both agriculture and health policy. In 1936 this was followed by a national diet exhibition. The 1930s were also notable for the introduction of the so-called 'Oslo breakfast' for school children.[4]

The war years, 1940–45, again resulted in food scarcity, especially in the cities, and it was essential that assuring food security and healthy nutrition was on the agenda of the social democratic governments ruling the country in the postwar period. The Norwegian Nutrition Council was given an important role in defining and implementing a healthy national diet, and a major milestone in the development of Norwegian nutrition policy was the publication of a White Paper on food and nutrition policy in the mid-1970s.[5] The focus in this policy was on enough food and energy to ensure good nutrition, and the approach was characterized by a high level of public regulation, paternalism and consensus. The policy was renewed in 1980–81, when it still emphasized food supply and food security.[6] The next White Paper of 1992–93 shifted the focus to diet-related health and social inequalities in health and these topics remain the main objectives today.[7]

---

3   Meltzer and Nordhagen 2007: 73–90.
4   Lyngø 2007: 27–39.
5   Ministry of Agriculture 1975.
6   Ministry of Health and Social Affairs 1981.
7   Ministry of Health and Social Affairs 1993.

Excess Bodyweight in Norway

From 1945 to 1985 Norway went through an extraordinary transition from one of the poorest countries in Europe to one of the richest. After the war, cardiovascular diseases increased and, at the same time, new knowledge about the quality of the diet, and especially the effects of saturated fat intake, became available. In 1975–6, nutrition policy identified being overweight as a risk factor for cardiovascular diseases and diabetes. Being overweight or obese was said to have increased in the 1950s and 1960s but to have stabilized in the mid-1970s.[8] The period of Norwegian affluence dates from 1985. This coincided with an inclusion of more fruits and vegetables in the diet but, at the same time, there was a spread of products rich in fat and sugar and consequent increases in being overweight. The weight of 40 year-old men in Norway increased by 9 kg [19 lb] over the past 30 years and correspondingly women have gained almost 4 kg [9 lb].[9] During this period health has been viewed more as an individual responsibility and there has been a focus on giving consumer information as a strategy to improve personal health.

The Historic Highlights of Food Labelling

Nineteenth-century developments in chemistry were important, both for the growth of food manufacturing and the establishment of state control.[10] The expansion in Norwegian industrial food production, coupled with a surge in urbanization which came at the end of the century, posed challenges for food safety, and there was a obvious need for consumer protection. Early in the twentieth century, the adulteration of food (for example, adding copper to make pickles green) was a health concern and Norwegian health campaigners began looking for best practice abroad.[11] For example, the United States Congress passed the Pure Food and Drug Act of 1906 to ensure food safety.[12] This regulated labels so that they could not contain false or misleading statements about ingredients, and required a clear statement of the food contained inside the package. In 1928 the US Food and Drugs Administration (FDA) added net weight, and the names and addresses of food processors to the list of required information. After 1958 all food labels had to list additives, and by the 1970s it was mandatory to list basic nutritional information. The US Nutrition Labelling and Education Act was implemented in June 1994. Manufacturers were then required to list ingredients in descending order of predominance by weight, and there were five other major changes: larger

---

8   Ministry of Health and Social Affairs 1981.
9   Tverdal 2001: 667–72.
10   The first food control system in Europe was in 1856 in Brussels. For details see Scholliers ICREFH IX 2007.
11   Elvebakken 2001.
12   U.S. Food and Drug Administration 1993.

type, nutrition facts, per cent daily value, fat, cholesterol, fibre, sugar, calories from fat, and the more realistic computation of serving sizes.

In Norway an act was proposed on food control in 1911 arising out of concern for public health and the rights of consumers and retailers, but legislation was not passed until 1933 because of conflicts between health authorities and trade and agricultural interest groups. Industrial interests in particular were opposed to interference with the free market.[13] Food scandals (e.g. Bovine Spongiform Encephalopathy or BSE, salmonella) at the end of the twentieth century brought about a new awareness of risk, safety and trust in food.[14] Today food labelling, in addition to safety assurance, also aims at protecting consumers' rights and providing information on nutritional quality and health advice. Pressure is growing for clearer labelling that makes it easier for consumers to understand labels and compare products. The current system of food labelling in Norway follows the basic principles of the European Union (EU) legislation which aims to ensure fair competition among producers by standardization of nutrition labelling to increase consumers' access to information, and reduce risks to individual consumer's safety and health.[15] Norway is not a member of the EU but, through being a member of the European Economic Area, she has agreed to enact legislation in the area of consumer protection similar to that passed in the EU. The first EU labelling Directive was passed in 1997, and this was replaced in 2000 by a new General Labelling Directive 2000/13/EC, which has been amended several times, for instance to include allergen labelling. In addition, there has been related legislation on sweeteners, additives, packaging gas, genetically modified organisms, novel foods, and other topics. Labelling on pre-packaged foodstuffs must now include: name, list of ingredients, quantity, potential allergens, the minimum durability date and keeping conditions.

In Norway, according to the nutrition labelling Directive of 1990 (with later amendments), nutrition labelling was optional but became compulsory if a nutrition claim appeared on the label or in advertising. The first suggestion for a new EU Directive on nutrition and health claims was proposed in 2003 and Regulation 1924/2006 was implemented with effect from 1 July 2007. This aims at protecting consumers' health and rights, and rebuilding consumer confidence in food safety. According to the Regulation, nutrition and health claims which encourage consumers to purchase a product are prohibited if they are false, misleading or not scientifically proven.

---

13   Elvebakken 2001.

14   Kjærnes, Harvey and Warde 2007.

15   Differences between national laws on labelling can lead to unequal conditions of competition.

Simplified Nutritional Labelling

Simplified labelling, signposting and logos (e.g. environmental, organic, fair trade) started appearing on packages in the 1980s. For example, the white swan, a Nordic logo for environmentally friendly products, was introduced in 1989. In the same year Sweden and Australia introduced simplified nutrition labelling. In Sweden, the authorities issued rules for a keyhole logo for foods with low content of fat, sugar and salt or high content of dietary fibres; in Australia, it was the heart foundation that started the 'pick the tick' campaign. New Zealand followed Australia and implemented a similar programme in 1991. A heart mark was also introduced in 2000 in Finland by the heart and diabetic associations. In the United Kingdom there is an ongoing debate about the voluntary traffic-light labelling (green – eat plenty; orange – eat in moderation; red – eat sparingly) recommended by the authorities. In response, businesses have introduced their own signposting system showing percentage of Guideline Daily Amounts for different nutrients.

Today simplified nutrition labelling exists mainly due to national requirements, but there are both Nordic and European initiatives. The Bureau Européen des Unions de Consommateurs (BEUC), the European consumers' organization, supports the development of a simplified labelling scheme and has initiated a project to try to develop a consensus amongst interested parties (consumers, industry, retail, public health experts) on an EU-wide model for providing government-endorsed nutritional information on packs in a simplified form.[16] BEUC has developed guidelines for a signposting system and recommends the use of a nutritional analysis table, including information in the form of Recommended Daily Amounts, combined with simplified front-of-pack signposting that conveys certain essential information in a manner that facilitates consumer choice at the point of purchase.

Simplified labelling of foods is today promoted in Norway as a new public health tool for assisting consumers to identify easily which foods are healthy options and which are high in energy, fat, sugar, salt or fibre. Non-governmental organizations (NGOs) and food retailers have taken the first steps to develop simple ways of communicating nutritional information to consumers. A simplified labelling system (the Swedish key hole) was introduced in parts of the Norwegian grocery sector in 2006 and the Norwegian authorities have now decided to implement simplified nutritional labelling in 2009.

Consumers' Views

Food choice is a complex consumer practice that has been described as based on routines, rationality and intuition.[17] Labelling may play some role in providing

---

16   Available at: http://www.beuc.org [accessed: 24 February 2009].
17   Berg 2005.

information for choices based on rationality. Labelling aims at informing and protecting the consumer, but overloading complex information on to food products may also result in misunderstanding and misinterpretation. More information is not always better for consumers, and a simplified food labelling scheme may be a way to enable a broader spectrum of consumers to identify easily which foods are healthy options and which are high in energy, fat, sugar, salt or fibre. A recent review of consumers and nutrition labelling showed that labelling has a limited but important role in promoting healthy diet.[18] Earlier research on food labelling has shown that consumers are interested in nutrition information but that they do not always read and understand food labels.[19] Based on research in the United Kingdom, voluntary labelling schemes are more likely to confuse and mislead consumers rather than inform them. Consumers do not always understand what the labels and logos mean.

To get an understanding of how Norwegians consumers view food labels and especially simplified nutritional labelling, a survey was conducted in 2007.[20] The target was the population between 18–80 years of age and a sample consisting of 1,000 respondents was interviewed by telephone. The survey revealed that when presented with a list of possible information on food labels, Norwegian consumers in general look at product information (date, price, brand name, ingredients) and less often at nutritional information. About half of the respondents reported that they usually look at fat, sugar and additives. However, many claimed that they choose their daily diet on the basis of what they think is healthy. There may be various reasons for consumers reporting that they do not very often look at nutritional information. The survey showed that many Norwegian consumers are content with current food labelling. It may be that they are not very interested in nutritional information because they think they already have a healthy diet and that it is fairly easy to find healthy foods in the grocery store. Alternatively, experience may have shown that it is not easy to find the information they want on the package and thus they do not look for it.

The majority of participants were in favour of simplified nutritional labelling and think that making healthy choices would be easier if it were introduced. One in four reported that they had seen or heard about labelling foods by symbols in Norwegian grocery stores. Many said that they would prefer a traffic-light type of label but this is somewhat surprising considering that so many also said that they do not at present read nutritional information on the label. This may be a methodological issue – people usually tend to say that they wish to have as much information as possible. But it may also reflect that a traffic-light type of label fits better with people's perceptions of the healthiness of foods. Consumers seem to have a somewhat ambivalent view of simplified nutrition labelling. They are

---

18   Cowburn and Stockley 2005: 21–8.

19   Grunert 2002: 275–85, Wandel 1997: 212–9, Wang et al. 1995: 368–80, Cowburn and Stockley 2005: 21–8.

20   Roos 2007.

mostly positive, but it remains unclear how much they would really use it when choosing foods in the grocery store, and whether it really would motivate them and make it easier to make choices. It is important to consider that health information is only one of many potential factors consumers take into consideration because food choice is a complex process influenced by both social processes and individual decision-making.[21]

Government and Policy-Makers' Views

One of the major health challenges facing governments and health-policy makers today is to find successful ways to modify food choice.[22] Public-health policy is underpinned by the notion that people should take responsibility for their own health. In the area of food this means that policies have to depend on consumers' food choices, although there are critics who claim that the assumption that there are rational consumers who see food primarily as a means to health actually results in a reduction in consumers' autonomy. If we treat food as a medicine we may neglect its important roles in social bonding, and in shaping individual and group identities.[23]

Health-policy makers today focus on motivating people to make food and other lifestyle choices that are better for their health and prevent obesity. This means that the success of policies comes to rest heavily on consumer behaviour. Empowerment, choice and responsibility are central themes. Information and labelling receive much attention, but people should not have to choose health in order to eat healthily. Labelling is useful but its limitation is that this type of information at point-of-purchase probably only benefits those who read labels and are motivated to change.[24]

Industry's and Retailers' Views

Competitiveness in markets today has become a key issue for the food sector. The food chain is said to have changed from a supply system to a demand system and, as a result, producers increasingly consider what their customers want. However, consumers' food choices are shaped and largely determined by the producers, processors and retailers. In particular, corporate concentration in the food-supply chain has resulted in power accruing to a few large global companies.[25]

---

21   Kjærnes and Holm 2007.
22   Buttriss et al. 2004: 333–43.
23   Food Ethics Council 2005.
24   Buttriss et al. 2004: 333–43.
25   Lang and Heasman 2004.

Consumers are said to worry about the health and safety of food production and to question the ability of the modern food system to provide safe food.[26] Information and transparency have become part of the strategies used by large retail chains in their competition for market share. Food manufacturers and retailers undertake some of the regulatory work previously done by national governments because self-regulation is a way to add value and reduce liability. They need to add value to food in order to survive in the marketplace and, because little money is made by selling fresh fruits and vegetables that form the mainstay of healthy eating advice, they turn to processing because it enables the creation of new qualities. Health-related labelling is then a means of locking in the claims they wish to make about high-price products.

Health-related food labels may also serve as bench marks highlighting and potentially rewarding health-related aspects in the product development process. Where manufacturers and retailers improve the nutrient profile of foods, labelling may have positive effects for consumers without them having to actually read the messages. For example, in Australia and New Zealand the salt content in some foods (breakfast cereals, bread, and margarine) was reduced in the late 1990s after the introduction of the tick logo.[27]

It is possible that health-related food labelling may move the focus away from diets and lifestyles, towards the qualities of single industrial products, thereby also taking attention away from healthier fresh produce. There is also an objection that in technologically and otherwise dynamic fields like food production, labelling may hamper product developments as 'good enough' solutions are rewarded on an equal footing with those that search to optimize the healthiness of their product. Businesses have been criticized for their focus on differentiating products rather than raising base-line standards of nutrition.[28]

Conclusions

The obesity epidemic has been referred to as a driving force for simplified nutritional labelling in Norway. NGOs and retailers were the first to put this on the agenda, but now the authorities have decided to implement a form of simplified nutritional labelling. Many interested parties are involved in the patterns and processes behind food labelling. Marion Nestlé has described the interactions in the United States at the beginning of the twentieth century, when industry, regulatory agencies, the public, and Congress established a cyclical pattern. Manufacturers marketed products with health claims; the FDA responded with regulations; the marketers filed objections; the courts ruled in favour of the food industry; there

---

26   Henneberry et al. 1998: 83–94, Macfarlane 2002: 65–80.
27   Williams et al. 2003: 51–6, Young and Swinburn 2002: 13–19.
28   Food Ethics Council 2005.

was lobbying; the Congress passed laws limiting FDA authority; and the marketers took advantage.[29]

To get an understanding of why health-related food labelling today has received more attention and support, it is useful to explore similarities and differences in the views expressed in a wider context. The state, consumers and businesses have very different roles in the food system. The government is mainly a regulator and policy maker, consumers are involved in everyday food-choice processes, and companies provide goods and services. The power of governments has been declining in the food system as corporate concentration has given big companies more power.[30] Different interests also have different roles and expectations in relation to food, and refer to food in different ways. Food is mainly presented as natural or for its nutritional values by health-policy makers, whereas consumers see it more in cultural terms as part of social relations and everyday life; while retailers and big companies present food merely as a commodity for sale and purchase.[31]

Health is a central value in modern Western societies and food is very closely linked to various aspects of health. The government and authorities are concerned about public health (currently much focused on excess bodyweight and obesity) and the costs of health care. By framing being overweight or obese as an epidemic, food may be equated to a 'medicine'. The role of health experts has come to empower people and motivate and support them to improve their own health, make healthy choices, or take their 'medicine'. In consequence, health again becomes more a matter of individual responsibility. Health policy seems to assume that people will choose food based on health. People learn as children to make a distinction between healthy and unhealthy foods, good foods and bad foods.[32] During the past decades the link between food and health has become internalized as part of everyday eating, and health concerns are incorporated in public debate and writings about food.[33] Consumers also report that they are interested in healthy food.[34] But health is not often foremost in people's minds when they decide what to eat. Food is for many mainly associated with taste and pleasure, which is often seen as an antithesis of healthy food.[35]

The food industry, in line with governments, seems to prefer individualistic approaches to health. Businesses thus also treat food like 'medicine' and invite consumers to improve their own health. But their remit is not to improve public health, and large food companies are more concerned with market saturation and the limits to growth that this presents for their businesses, while at the same time facing the risk that they might be held legally and morally liable for sales of their

---

29   Nestlé 2002.
30   Lang and Heasman 2004.
31   Jacobsen 2004.
32   Johansson et al. 2009.
33   Holm 2003, Warde 1997.
34   Torjusen 2004, Roos 2006.
35   Warde 1997, Makela 2002.

products. Selling more food has become difficult and one response for companies is to differentiate products, charging more for foods with health qualities.

Different interest groups have different roles in health-related food labelling. Food-labelling legislation was developed for food-safety reasons by governments. The new EU Regulation on nutrition and health claims shows that governments can have an active role in regulating the health-related information included on food labels. However, it is relevant to question if labelling is enough to achieve the policy goal of enabling people to improve their own health. It builds on an expectation that people have the capacity to do more and that the state accepts that information included on labels will boost the demand for healthy food. Labelling has a potential to help improve health but it is not enough. By focusing on consumers' choice and decisions, governments underplay the importance of supply-side decisions and other policy measures that could make a difference to people's health. There is an over-optimistic expectation that the market will respond to demand and reduce salt, harmful fats, and sugars in processed food.

Consumers can use their choices and political activities to influence what is on food labels. However, as shown in this chapter, consumers have not been very active using or requesting simplified nutritional labelling in Norway. Simplified nutrition labelling may help consumers choose healthy products in the grocery store, and may have a positive health outcome for consumers if it induces producers to change the composition of their products to fulfil criteria for labelling.

Companies select health-related information for marketing and advertising, when the benefits to them of providing the information outweigh the cost. Health-related labelling may also provide incentives for manufacturers to change the composition of older products and develop new products that promote health. However, there is also a risk that this may lead to an increased use of food fortification and health-related marketing to boost sales of products that should be eaten only sparingly according to health advice.

Health-related food labelling fits well with the idea of an 'obesity epidemic' and the view of food as 'medicine'. There is an overlap between the ways that health-policy makers, the food industry and food retailers refer to health-related food labelling which may strengthen its implementation. Consumers have a wider view of food and are more ambivalent about the role of this type of food labelling. Consumers tend to support simplified nutritional labelling but they are not active advocates for it.

Norway has moved from undernutrition to overnutrition in the twentieth century as nutrition policy has shifted from focusing on supply and assuring enough food to the population to focusing on health and individual responsibility.

References

Berg, L. 'Tillitens Triagler: Om Forbrukertillit og Matsikkerhet' [The Triangle of Trust: Consumer Trust and Food Safety], *SIFO Fagrapport* No. 1, Oslo, 2005.

Buttriss, J., Stanner, S., McKevith, B., Nugent, A.P., Kelly, C., Phillips, F. and Theobald, H.E. 'Successful Ways to Modify Food Choice: Lessons from the Literature', *Nutrition Bulletin* 29, 2004.

Coveney, J. *Food, Morals and Meaning: the Pleasure and Anxiety of Eating*, London, 2000.

Cowburn, G. and Stockley, L. 'Consumer Understanding and Use of Nutrition Labelling: a Systematic Review', *Public Health Nutrition* 8, 2005.

Elvebakken, K.T. 'Næringsmiddelkontroll – Mellom Helse – og Næringshensyn' [Food Control – for Health and Honest Trade], *Tidsskrift for Den Norske Lægeforening* 121, 2001, 3613–6.

Food Ethics Council 'Getting Personal: Shifting Responsibilities of Dietary Health', 2005. Available at: http://www.foodethicscouncil.org/node/115 [accessed 19 April 2009].

Grunert, K.G. 'Current Issues in the Understanding of Consumer Food Choice', *Trends in Food Science & Technology* 13, 2002.

Handlingsplan for Bedre Kosthold i Befolkningen (2007–2011), Oppskrift for et Sunnere Kosthold [Action Plan: Prescription for a Healthier Diet], Oslo, 2007.

Henneberry, S.R., Qiang, H. and Cuperus, G.W. 'An examination of Food Safety Issues', *Journal of Food Products Marketing* 5, 1998.

Holm, L. 'Begreper om Mat og Sundhed' [Concepts of Food and Health], in Holm, L. (ed.), *Mad, Mennesker og Måltider – Samfundsvitenskabelige Perspektiver*, Oslo, 2003.

Jacobsen, E. 'The Rhetoric of Food: Food as Nature, Commodity and Culture', in Lien, M.E. and Nerlich, B. (eds), *The Politics of Food*, Oxford, 2004.

Johansson, B., Makela, J., Hillén, S., Roos, G., Jensen, T.M., Hansen, G.L., and Huotilainen, A. 'Nordic Children's Foodscapes: Images and Reflections', *Food, Culture & Society* 12, 2009.

Kjærnes, U., Harvey, M. and Warde, A. *Trust in Food. A Comparative and Institutional Analysis*, Basingstoke, 2007.

Kjærnes, U. and Holm, L. 'Social Factors and Food Choice: Consumption as Practice,' in Frewer, L. and Trijp, H. van (eds), *Understanding Consumers of Food Products*, Cambridge, 2007.

Lang, T. and Heasman, M. *Food Wars: the Global Battle for Mouths, Minds and Markets*, London, 2004.

Lyngø, I.J. 'Et Melkedrikkende Folk – Melkens Nye Status i Mellomkrigstidens Norge' [Milk Drinking People – the New Status of Milk in Norway in the Time between the World Wars], *Arr Idéhistorisk Tidsskrift* 19, 2007.

Macfarlane, R. 'Integrating the Consumer Interest in Food Safety: The Role of Science and Other Factors', *Food Policy* 27, 2002.

Makela, J. *Syomisen Rakenne ja Kulttuurinen Vaihtelu* [The Structure of Eating and Cultural Variation], Helsinki, 2002.

Meltzer, H.M. and Nordhagen, R. 'Norsk Matkultur i et Helseperspektiv' (Norwegian Food Culture in a Health Perspective), in Amilien, V. and Krogh, E. (eds) *Den Kultiverte Maten: en Bok om Norsk Mat, Kultur og Matkultur*, Oslo, 2007.

Nestlé, M. *Food Politics*, Berkeley, 2002.

Oddy, D.J. 'Food Quality in London and the Rise of the Public Analyst, 1870–1939', in ICREFH IX, 2007, 91–103.

Roos, G. *Kropp, Slanking og Forbruk: SIFO-Survey Hurtigstatistikk* [Body, Dieting and Consumption: SIFO-Survey Statistics], Oslo, 2006.

Roos, G. *Symbolmerking av Sunn Mat: Forbrukersurvey* [Symbol Labelling of Food: a Consumer Survey], Oslo, 2007.

Scholliers, P. 'Food Fraud and the Big City: Brussels' Responses to Food Anxieties in the Nineteenth Century', in ICREFH IX, 2007, 77–90.

Torjusen, H. 'Tillit til Mat i det Norske Markedet: Hvordan Oppfatter Forbrukerer Trygg Mat?' [Trust in Food in the Norwegian Market. How do Consumers View Food Safety?], *SIFO Oppdragsrapport* no. 11, Oslo, 2004.

Tverdal, A. 'Prevalence of Obesity among Persons Aged 40–42 Years in Two Periods', *Tidsskr Nor Lægeforen* 121, 2001.

United States Food and Drug Administration 'Good Reading for Good Eating', *FDA Consumer* May 1993. Available at: http://www.fda.gov/fdac/special/foodlabel/goodread.html [accessed 19 April 2009].

Wandel, M. 'Food Labelling from a Consumer Perspective', *British Food Journal* 99, 1997.

Wang, G., Fletcher, S.M. and Carley, D.H. 'Consumer Utilization of Food Labelling as a Source of Nutrition Information', *Journal of Consumer Affairs* 29, 1995.

Warde, A. *Consumption, Food and Taste*, London, 1997.

WHO, *Obesity – Preventing and Managing the Global Epidemic, Report of a WHO Consultation on Obesity, Geneva, 3–5 June, 1997*, WHO/NUT/NDC/98.1, Geneva, 1997.

Williams, P., McMahon, A. and Bousted, R. 'A Case Study of Sodium Reduction in Breakfast Cereals and the Impact of the Pick the Tick Food Information Programme in Australia', *Health Promotion International* 18, 2003.

Young, L. and Swinburn, B. 'Impact of the Pick the Tick Food Information Programme on Salt Content of Food in New Zealand', *Health Promotion International* 17, 2002.

Chapter 9

Sugar Production and Consumption in France in the Twentieth Century

Alain Drouard

Introduction

Sugar[1] is a product charged with history.[2] This chapter seeks to analyse not only the changes of status which accompany the rise of its consumption but also the debates that have been caused in contemporary France for its part in the development of obesity.

France was a pioneer in the beet sugar industry in the nineteenth century. In the twentieth century France became the eighth biggest producer of sugar in the world and the largest producer of beet sugar, while being one of a few countries to produce both cane and beet sugar. France was the largest European producer of sugar if one considers both the production of metropolitan France and its Overseas Departments of Réunion, Guadeloupe, and Martinique. This result was reached thanks to the modernization and centralization of sugar production. Indeed, over the half century after the Second World War, the number of growers was reduced by two-thirds, the surface cultivated per grower tripled and production multiplied by ten. Each factory was nine times larger but at the same time the expenditure of energy was down by two-thirds and manpower by four-fifths. There were some sixty sugar refineries at the beginning of the 1980s; today there are no more than about thirty producing the same amount of sugar – about four million tons – which shows important gains in productivity.

For a long time, France consumed less sugar than Britain or the United States but in the twentieth century domestic consumption in France progressed very quickly. It nearly doubled to exceed 30 kg per annum before stagnating at this level for the last thirty years. Sugar consumption therefore changed: from a rare and expensive product to become a commodity of widespread human consumption. However, over the last thirty five years consumption has changed markedly: the

---

1 The term sugar in the singular refers to sucrose in French regulations and context and has the chemical formula $C_{12}H_{22}O_{11}$. It is extracted from cane or beet by industrial processes of refining. The simple sugars, glucose, fructose and galactose are termed monosaccharides while sucrose, maltose and lactose are disaccharides. Complex carbohydrates (CHOs) are oligo- or polysaccharides, e.g. starch.

2 Mintz 1985.

domestic consumption of sugar (as sugar) now represents no more than 20 per cent of total consumption while nearly 80 per cent is indirect consumption, primarily by the food industry.

Beyond the discontinuities which the two world wars brought, the evolution of the production, consumption and use of sugar has varied over the years. Before 1945 production and consumption progressed slowly; after 1945 they increased and the uses of sugar changed, which did not fail to generate numerous questions about the nature and the properties of the product. By the end of the twentieth century, because of the development of obesity in France typical of developed countries, sugar became seen as a problem of public health which reactivated the debates between the 'saccharophobes' and the 'saccharophiles'.

From Sugar as a Condiment to a Food and Ingredient

In 1826, Brillat-Savarin observed that 'the use of sugar has become more frequent each day, more general and there is no food substance that has undergone more amalgamation and transformation...' He specified its uses: pure sugar mixed with water made a 'healthy, pleasant and sometimes salutary drink, like a remedy'. It was used to manufacture syrups, ice creams, pastries, jams, and liquors, while 'mixed with coffee, it brings out its flavour'. Brillat-Savarin added: 'The use of sugar is not restricted to that. We can call it the universal condiment that does not spoil anything ... Its uses vary *ad infinitum* because they depend on peoples and individuals'.[3]

Because of its increased availability at the end of the nineteenth century sugar was no longer merely a remedy or a condiment.[4] While it was consumed more frequently with coffee or breakfast's *café au lait*, its use in cooking and pastry making had also spread. It was used in the preparation of jams as well as desserts, ice creams and sorbets. However, it was eaten only in small quantities, mainly by children and women:

> If the French housewife had not been forced, by the enormous burden of the tax, always to lock up sugar like a luxury article, to restrict the use of it, perhaps we would consume it today less than England but certainly as much as the United States...
>
> If the housewife did not hide her sugar any more, her husband and her children would find in this food a healthy fulfilment of the kind of appetite that

---

3   Brillat-Savarin 1967: 88.

4   From 2,600 tons in 1827–8, production increased to 75,000 tons in 1852–3 and 805,000 tons in 1899–1900, while the price of refined sugar was divided by three between the beginning and the end of the nineteenth century in spite of a very heavy tax on it.

makes the tired worker wish for pleasant sensations like those which push him towards alcoholic beverages.[5]

To develop domestic consumption, sugar producers advertised the experiments of doctors and physiologists and from then on treated sugar as a source of energy, even a health food, an essential food, a food of *strength* especially useful for the working classes:

> Sugar, a hydrocarbonated food, is the best generator of strength: it repairs the loss of muscles and advantageously replaces alcohol, which is only a momentary stimulant and presents serious drawbacks for the body.[6]

According to experiments conducted by an English physiologist on people given a fixed diet, with or without sugar, work output increased by adding sugar to food.[7] Sugar as a food also became an ingredient used by industries making biscuits, sweets, chocolate, syrups and liquors. It was used in the wine production for chaptalization.[8]

Production

Sugar beet production in France did not cease, despite the competition of cane sugar; indeed, at the end of the nineteenth century the production of beet sugar exceeded that of cane. While France produced 75,000 tons of sugar at home in 1850, its production reached 450,000 tons in 1875, almost 700,000 tons at the end of the 1880s and it had become the leading European producer, followed by Germany. This increase in production was due to the extension of sugar beet cultivation in the North of France, improvements in productivity and quality, as well as to the technological progress and the centralization of sugar refineries. From 530 refineries in 1875, output rose to 1 million tons at the turn of the century. On the eve of World War I, 206 refineries produced 700,000 tons but after the

---

5   Hélot 1900: 180.

6   Hélot 1900.

7   Hélot 1900: 187. At the same time, Professor Brouardel wrote: 'This is why finally fats and carbohydrates, cane sugar in particular, considered separately in a state of absolute purity, must be proclaimed excellent foods, from the point of view of their special energy function, the production of strength'. Brouardel 1904: 388.

8   The sweetening of wine instituted by Jean-Antoine Chaptal when he was one of Napoléon's ministers, was regulated. The law of 29 July 1884 authorized a maximum of 20 kg of sugar for 3 hectolitres of wine. Chaptal wrote 'it is obvious you can reach the content of alcohol wanted whatever the quality of the must by adding more or less sugar'. Chaptal 1801: 91.

conflict the beet fields and the factories had been destroyed. At the end of the war in 1918 there were no more than 50 refineries producing 100,000 tons.

The rebuilding of the beet industry from 1919 onwards was accompanied by further centralization of production but the economic crisis of 1929 destroyed its plans. The sugar-producing countries signed agreements (the Chadbourn Plan in 1931, the London Agreement of 6 May 1937, which created the International Sugar Council) to try to control the world market surplus. While the crisis faded away at the end of the 1930s, World War II changed the situation. It was a time of restrictions and rationing, which necessitated rebuilding everything in 1945.

In 2004, Jean Airiau showed that in the aftermath of the war, the sugar economy in France rested on the interrelationship between three interested parties.[9] First, the growers who decided how much land to assign to beet cultivation, then the refineries, which decided what quantity of beets to buy, and finally the authorities who defined production quotas as well as the prices at which it was appropriate to buy beet and sell sugar.[10] The period between 1945 and 1953 witnessed an effort to maximize production, which was organized within the framework of first national then community quotas. The 'beet plan' of 1946 encouraged the resumption of sugar production, which also benefited the distillery sector while the need for alcohol instead of petrol in vehicles' carburettors decreased.

The production of alcohol from sugar exceeded peacetime needs and became expensive for the authorities. The sugar plan of 9 August 1953 thus instituted quotas for the production of sugar, the closing of many distilleries and the transformation of some of them into sugar refineries. Centralization developed during this period. The modernization of transport and factories went hand-in-hand with the use of new seeds, the progressive mechanization of sowing and harvesting, as well as the introduction of 'saccharimetry' to replace the measurement of density.[11]

With the introduction of the Common Market for sugar in 1968, French production continued to develop and reinforce its competitiveness thanks to the modernization of its production equipment. In the 1980s, French production ceased growing because of the stagnation of consumption and of the weakness of the world prices but production technology was increasingly powerful and productivity had never been so strong. The number of sugar refineries was quickly reduced and they appeared to be functioning without workers. The production was more and more automated and computerized.

---

9   Airiau 2004: vol. 1, 15.

10   Right from the start the State played a major part in the sugar economy in France. After the Imperial decree of 15 January 1812 'regarding the manufacture of beet sugar' several laws (19 May 1860; 29 July 1884; 5 August 1890; 28 June 1891) were passed throughout the century to organize the production and the manufacture of sugar and to define the taxation on sugars.

11   Saccharimetry is assessing the concentration of sugar solutions.

Consumption

Industrial sugar refers to that used by the food industry (chocolate, biscuits confectionery manufacture and distilling and brewing) while the direct consumption of sugar refers to that bought in shops in weights of less than 10 kg, such as a 1 kg box of lump, caster or crystal sugar. French national consumption is calculated as the difference between production plus imports minus exports. The difference between national consumption and direct consumption is the amount used by industry or indirect sugar consumption. Apparent or individual consumption is obtained by dividing the amount of sugar produced or sold by the number of the population. However, this technique leads to an over-estimate of consumption since what is sold is not necessarily consumed because of losses and waste. Individual consumption is distinguished from the concept of availability based on production figures expressed in kg/head/year. Availability takes into account imports and exports and thus corresponds to the quantity of direct plus indirect consumption. At the international level, classifications of carbohydrates vary considerably, which is a source of confusion and difficulty when one seeks to compare the consumption in one country with another.

Stimulated by the fall in prices, and due to technical progress and international competition, sugar consumption increased throughout the nineteenth century. While consumption was only 3 kg per head per annum in 1837, it was six times that at the beginning of the twentieth century, taking into account the sugar of the Overseas Departments. Sales reached 600,000 tons in 1900 and 700,000 in 1913. Consumption was then 18 kg per head per annum and on the eve of the Second World War reached 25 kg per head before rationing imposed a limit of 6 kg per annum, with a few exceptions allowed 12 kg per annum.

In the aftermath of the war, sugar consumption experienced a remarkable development in France. According to the results of a survey of 2,685 people carried out by the National Institute of Demographic Studies in June 1948, the French hungered for sugar while they were still subjected to rationing:

> 90 per cent of the people complained about the insufficiency of the ration of sugar…One household out of two manages to get more sugar than the official ration…eight out of ten would accept a price increase for sugar to obtain an increase in the ration. Nine people out of ten favoured the production of sugar over that of alcohol…

The survey concluded:

> The conclusions which emerge from this survey are clear: the public, asked to choose between an increase in the ration of sugar, even for more money, and the status quo, do not hesitate. They accept the principle of an increased

production of sugar obtained at the expense of the manufacture of alcohol for direct consumption.[12]

From 496,000 tons during the Occupation, national consumption doubled to reach 950,000 tons during the marketing year 1949–50 and, at the end of the post-war boom in 1973 had doubled again to 2 million tons. Since the middle of the 1970s, the total consumption seems to have stabilized at this level, corresponding to consumption per head of approximately 35 kg per year. If one believes the information made public by the *Centre d'études du Sucre* (CEDUS), which refers to individual investigations of consumption going back to 2000, sugar consumption per head per year would be approximately 27 kg.[13] Beyond the total consumption, which is stable around 2 million tons, it is necessary to distinguish the evolution between direct and indirect consumption. Sugar consumed directly declined in some thirty years since the 1970s from more than 45 per cent to 20 per cent of the total market. This could not be explained only by the price.[14] Changing food consumption played an essential part here, for the French make fewer pastries and jams at home and resort more and more to buying fresh or industrial pastries. They buy more processed products and have decreased similarly their purchase of lump and powdered sugar.

In France, as in other countries of Western Europe, sugar consumption thus seems to have reached a level of saturation. Although sugar can be used in a great number of food-related and unrelated fields such as chemistry, biotechnology or wine production, it nevertheless faces more competition from other sweeteners – natural[15] or synthetic.[16] The consumption of these products, now authorized by European regulation,[17] is increasing because of their lower production cost and

---

12   Brésard 1948: 549–50.

13   The difference between the apparent consumption and consumption per head is explained by the fact that all the sugar sold is not completely consumed by the French population, because of losses and waste by the consumers as well as non-food uses. Besides, what is really consumed is not always declared in studies.

14   To buy 1 kg of lump sugar it was necessary for a consumer earning the minimum wage to work 1 hour 57 minutes in 1955, and 1 hour 14 minutes in 1987. However, the prices of lump and powdered sugar increased after 1990.

15   Natural sweetening substances include sucrose extracted from cane or beet, glucose with a slightly lower sweetening capacity found in many fruits but generally extracted from corn or potato by starch hydrolysis, fructose that has a strong sweetening capacity as in honey and isoglucose or HFCS (High Fructose Corn Syrup) which is generally extracted from plants containing starch (primarily maize).

16   The synthetic sweetening substances have a strong sweetening capacity and are amino acid compounds. The most recent chemical sweetening substance is aspartame, which gradually replaced saccharin and cyclamates. With the exception of saccharin, which is much older since it goes back to 1879, they all appeared during recent decades.

17   A 1902 law prohibited the use of saccharin for human consumption. This prohibition was then lifted by the European regulation on sweetening substances on 10

also because they are said to be energy free. The synthetic sweeteners are therefore more and more used in drinks because they do not have an effect on consistency while having a very high sweetening capacity. A new balance between sweetening products is developing in France and Europe.

A Recurring Debate and Health at Stake

In 1999 the World Health Organization (WHO) labelled obesity as a pandemic requiring immediate action.[18] It mobilized those in charge of public health at the international level. While France launched its National Programme Nutrition-Health in 2001, the French agency for medical safety of foods, the *Agence française de sécurité sanitaire des aliments* (AFSSA),[19] also tried to deal with the issue after assessing that in France 19 per cent of children and 41 per cent of adults were overweight or obese.[20] It is therefore not surprising that the current development of obesity in Europe revived old controversies on the properties of sugar between the pro-sugar or saccharophiles and anti-sugar or saccharophobes.[21] The attacks were particularly sharp at the beginning of the twentieth century when the naturopath, Dr Paul Carton, called sugar a 'dead food' and labelled it, with alcohol and meat, as fatal food:

> It is nothing more than an irritating drug, a denatured and concentrated chemical body whose contact is unphysiologic and aggressive. The work of incorporation of this dead chemical energy determines a deviation of the digestive processes

---

September 1994, authorizing as additives sugar-alcohols as well as aspartame, saccharin and some other products.

18   World Health Organization 2000.

19   AFSSA 2004. The term 'glucid' means carbohydrate (CHO).

20   Adult obesity is evaluated internationally by the BMI. (body mass index) corresponding to the ratio of weight expressed in kg over the square of height expressed in metres i.e. weight/size (in kg/m^2). The difficulty of measurement is shown by the discrepancy between the AFSSA figure and that published for France by the European Union. See Table 16.3.

21   In France, Joseph Du Chesne, King Henri IV's doctor, was one of the most violent saccharophobes of his time. In his *Diaeteticon Polyhistoricon*: 419, he warned against the toxic effects of sugar which creates thirst, overheats the bile and blackens the teeth and wrote: 'Under its whiteness, sugar hides a great blackness and under its softness an extremely large acrimony which equals that of strong water. So much so that it could dissolve and liquefy the sun itself.' In 1647, Garancières, a Frenchman who emigrated to England, violently attacked sugar: 'It is clear that sugar is not a food but an evil spell; that it is not a preservative but a destruction and that we should send it back to India, for before its discovery the consumption of the lungs was not known but was brought to us with the fruit of our labour.'

and a deep visceral devitalization, which is accompanied by hepatic and vascular congestion.[22]

Over the years, especially in recent decades, sugar has been accused of causing various evils from the plague to tuberculosis, scurvy to haemorrhoids – and freckles.[23]

In reaction to this criticism, the sugar producers could not help but react and counterattack. In 1932, they created CEDUS[24] as a professional body in the field of beet and cane sugar and a Trade Association whose vocation was 'to ensure the information, documentation and promotion of sugar'.[25] In 1984, the members of the beet and cane sugar association founded the Centre of Food Information to gather within the same structure 'the organization of promotion, advertising, and events'.

These organizations promoted memorable advertising campaigns carried out under the name of the Sugar Group: from the 'cascades of dominos' in the 1980s to the 2005 advertisement which alerted consumers on the nonsense of 'a world without sugar'. In October 2006, the advertising campaign aimed at informing the public and opinion leaders on the consequences of the reduction, or even of the removal of sugar in the composition of foodstuffs. With the slogan 'When one removes sugar, do you know what one puts in instead?' this campaign explained that sugar reduction inevitably implies the use of substitutes little known by the consumer and intended to compensate for the natural properties of sugar. Another impressive initiative, the Week of Taste, was launched in 1990 led by the food critic, Jean-Luc Petirenaud. Each year, with the support of the Ministry of National Education, professionals from the food industry give lessons in taste to thousands of primary school students. Since the beginning of this initiative, 1.5 million children have taken part in these workshops. The effects of these campaigns are not obvious. As Fischler pointed out:

> In fact, the one most striking feature of current attitudes is ambivalence...
> Sweetness is perceived as gratifying. It also somehow makes for emotional

---

22   Carton 1942: 21.

23   See Abramson and Pezet 1971, Frederick and Goodman 1969, Yudkin 1972, Dufty 1975, Starenkyj 1981.

24   The former names of this organization were: in 1932, Committee of Studies, Hygiene and Use of Agricultural Products, in 1951 the Committee of Studies, Information and Propaganda for the Use of Sugar and, in 1955, the Centre of Studies and Documentation for the Use of Sugar.

25   CEDUS's mission is the 'creation and diffusion of documentary or teaching multimedia tools; organization and/or participation in public or professional demonstrations; support scientific research and the dissemination of knowledge in the fields of human diet, nutrition and health; management and actualization of a multi-field data base intended for professionals in the field and for sugar consumers'. It contributes to 'the valorization of the image of sugar as well as to the recognition of the part played by sugar in a balanced diet'.

security. Through offering and sharing, sweets symbolise and help bring about feelings of togetherness, social bonding, festive activity. Yet at the same time, as studies reveal, sweetness is also associated with an obscure sense of danger. Sugar appears as an often unnecessary gratification, one in which one should not indiscriminately indulge. [26]

Finally, the professional association promoted or carried out studies and research on sugar, which investigated consumer behaviour as well as the image and representation of the product. In 1996, Sucres et Santé [*Sugars and Health*], a study carried out by the Professor of Human Nutrition, Gérard Debry, was made public. As a critical study of the data published in 4,500 publications, it came to the conclusion that there was no relationship between sugar consumption and a certain number of pathologies like atherosclerosis, high blood pressure or obesity:

> The data from extensive research on food consumption does not allow us to establish a correlation between the consumption of sucrose and body mass index, may it be in children or adults.[27]

As an extension to this study, a new publication by CEDUS came to the same conclusions in 2005:

> Obesity is a multifactorial disease where a genetic predisposition, a long-lasting imbalance between the contribution and the expenditure of energy and environmental factors intervene...Apart from contributing to an excessive total energy contribution, CHOs cannot be held responsible in the aetiology of obesity. Consumed in great quantities, they are preferentially oxidized, whereas lipids are primarily stocked...In France, research on food consumption highlighted a negative relationship in children and adults between CHO consumption and excess weight...[28]

Type 2 diabetes, the most frequent type, constitutes a problem of public health because of the risks of cardiovascular pathologies that it generates. However, a largely widespread belief bound its occurrence with excessive CHO consumption, in particular of sucrose. Calling upon scientific literature and the experts of the French Agency of Medical Safety, CEDUS claims that it is not possible to connect clearly 'simple CHO contribution and type 2 diabetes'.[29]

If the conclusions of the report of AFFSSA *Glucids and Health* are similar, they are far less peremptory:

---

26   Fischler 1987: 89.
27   Debry 1997: 31
28   *Sucres et Santé* 2005: 35–6.
29   *Sucres et Santé* 2005: 38.

In spite of the analysis of a great number of studies, it is difficult to establish a clear connection between the consumption of simple CHOs (total or taken separately) and the incidence or the development of obesity, diabetes, cardiovascular diseases and cancers in adults. When harmful effects were observed with pathology, it was not possible to directly blame the nutrient or CHOs. The unmatched results of the various studies are probably linked to the fact that the populations studied and the methodologies used were not always comparable and that, as we have already stressed, the methodology used in these numerous studies can be questioned. The available data does not allow us to blame simple CHO consumption for the development of the pathologies in question but it does not make it possible to undermine it either.[30]

Regarding caries, CEDUS and AFFSSA underline the absence of any relationship between the level of sugar consumption and the prevalence of this pathology. Basing itself on the results of international research, the White Book of CEDUS explained in 2005:

A link between simple CHO consumption, especially sucrose, and caries is often suggested. One observes a fall in the number of caries in developed countries, mainly in Europe, for the past 30 years while the average availability of sugar in European countries is between 30 and 45 kg per capita per annum.

In addition, experimental data clearly showed, and had been showing for a long time, that the frequency of contact between the tooth and CHOs mattered more than the consumed quantity of this food.[31]

Whatever uncertainties and difficulties of interpretation the results of the research brought, the stakes of public health are such in France, as in the other European Union countries, that the authorities could not remain indifferent nor let CEDUS address the issue on its own. The National Programme Nutrition-Health (PNNS) launched in 2001 therefore recommended a reduction of 25 per cent of simple sugars while hoping it will reduce the risk of obesity.

As for AFSSA while it recommends limiting energy intakes and increasing energy expenditure for all age groups, it defined its position on carbohydrates:

Regarding carbohydrates, the group did not find any reason to question the recommended nutritional contributions for the general population and recommends a total CHO contribution of 50 to 55 per cent of the total energy contribution. This objective is seldom achieved in the French nutritional studies and the contribution must therefore be increased, in the form of complex CHOs.

---

30   AFSSA, *Rapport* 2004: 82.
31   *Le Livre Blanc Sucres et Santé* 2005: 44.

For simple CHOs, the group did not consider it necessary to provide a limit of consumption so as to not penalize foods like unsweetened dairy products or fruits and vegetables which have their own nutritional quality. Moreover, the research group recognizes the difficulty of fixing a quantitative limit to the consumption of simple added CHOs based on solid scientific ground. On the other hand, the simple added CHOs, whose consumption is constantly increasing and excess of which has proven to have toxic effects, must be reduced. The objective of the PNNS (National Programme Nutrition-Health) over five years to reduce by 25 per cent the population's consumption of simple CHOs must concentrate on simple added CHOs.[32]

The participation of the food industry is essential to reduce the 'passive' consumption of simple added CHOs. There are in particular many sweet food products whose content in simple CHOs could be reduced, without altering unfavourably the texture or any other functional property of the food. Food for infants, whose tastes are being formed, is particularly targeted.

Consumption of CHOs, little or not refined, (whole-wheat products, legumes, fruits and vegetables) must be strongly encouraged. The field of bakery-pastry should increase the share of products manufactured from whole-wheat flour in their products containing yeast.[33]

Conclusion

In France, sugar consumption increased until the 1970s. Since then it has stabilized yet has been transformed. The drop in sugar consumption at the table in households was compensated for by the increase in the use of sucrose as an ingredient in sweet foods, the consumption of which has been increasing (in sodas and sparkling beverages, sweets, dairy products, biscuits), while other simple CHOs like glucose or isoglucose are used more and more by the food industry which causes an increase in their consumption.

These changes, as well as the development of obesity, have revived the old polemic between 'saccharophiles' and 'saccharophobes' and raised a number of difficult questions at the same time. As shown by recent studies the obesity epidemic in France, as in the USA, is a much more complex issue than one thought in as much as there are many other contributors to the increase in obesity than food consumption or reduction in physical activity.[34]

Concerning the consumers, even if the majority of them do not or barely read the labels on the products they buy, it is necessary to note that the advertisements which are their main source of information are more numerous yet often contradictory. Confusion is all the more intense as nutrition science is still far from being able to

---

32   CNERNA-CNRS-AFSSA 2001.

33   AFSSA, *Rapport* 2004: 103

34   See Keith et al. 2006: 1–10.

bring reliable answers to questions one might have about the relationship between sugar and health.

References

Abrahamson, E. and Pezet, A.W., *Body, Mind and Sugar*, New York, 1971.

AFSSA, *Rapport 'Glucides et Santé: Etat des Lieux, Évaluation et Recommandations'* [Report 'Carbohydrates and Health: Inventory, Evaluation and Recommendations'], October 2004.

Airiau, J. *Le Sucre en France de 1945 à 1995* (De la Libération à la Libéralisation du marché) [*Sugar in France from 1945 to 1995* (From the Liberation to the Liberalization of the World Market)], unpublished thesis, University of Paris IV, 2004, 4 vols.

Brésard, M. 'L'Opinion du Public sur la Consommation de Sucre', [Public Opinion on Sugar Consumption'], *Population*, 3, July–September, 1948, 549–50.

Brillat-Savarin, J. *Physiologie du Goût, ou Méditations de Gastronomie Transcendante; Ouvrage Théorique, Historique et à l'Ordre du Jour, Dédié aux Gastronomes Parisiens, par un Professeur, Membre de Plusieurs Sociétés Littéraires et Savantes.* [A Physiology of Taste], Paris, 1967 (1st edition 1826).

Brouardel, P. 'La Saccharine: Etat de la Question du Point de Vue de l'Hygiène Alimentaire', [Saccharin: Status of the Issue from the Point of View of Food Hygiene], *Annales d'Hygiène Publique et de Médecine Légale* 4th series, 1, 5, 1904, 385–406.

Carton, P. *Les Trois Aliments Meurtriers* [The Three Fatal Foods], 1942 (1st edition 1912).

CEDUS, *Le Livre Blanc, Sucres et Santé* [The White Book, Sugars and Health], Paris, 2005.

Chaptal, J. *Traité Théorique et Pratique sur la Culture de la Vigne avec l'Art de Faire le Vin, les Eaux-de-Vie, Esprit de Vin, Vinaigres Simples et Compris* [The Art of Making Wine] Paris, 1801.

CNERNA-CNRS-AFSSA, *Apports Nutritionnels Conseillés pour la Population Française* [Contributions to Nutritional Advice for the French Population], Third edition, Martin, A. (ed.), Techniques and Documentation, 2001.

Debry, G. *Sucre et Santé Conclusions* [Sugar and Health Conclusions], Montrouge, 1997.

Du Chesne, J. *Diaeteticon Polyhistoricon*, Paris, 1606.

Dufty, W. *Sugar Blues*, New York, 1975.

Fischler, C. 'Attitudes towards Sugar and Sweetness in Historical and Social Perspective' in Dobbing, J. (ed.), *International Life Sciences Institute Human Nutrition Reviews: Sweetness* Berlin, 1987, 83–98.

Frederick, C. and Goodman, H. *Low Blood Sugar and You*, New York, 1969.

Garencières, De T. *Angliae Flagellum seu Tabes Anglica,* London, 1647.

Hélot, J. *Le Sucre de Betterave en France de 1800 à 1900* [Sugar Beet in France from 1800 to 1900], Cambrai, 1900.

Keith S.W., Redden, D.T., Katzmarzyk, P.T. et al. 'Putative Contributors to the Secular increase in Obesity: Exploring the Roads less Travelled', *International Journal of Obesity* 30, 2006, 1–10.

Mintz, S. *Sweetness and Power: The Place of Sugar in Modern History*, New York, 1985.

Starenkyj, D. *Le Mal du Sucre* [The Evil of Sugar], Québec, 1981.

World Health Organization *Obesity: Preventing and Managing the Total Epidemic*, WHO Technical Report Series, no. 894, Geneva, 2000.

Yudkin, J. *Sweet and Dangerous*, New York, 1972.

Chapter 10
Controlling Fat and Sugar in the Norwegian Welfare State

Unni Kjærnes and Runar Døving

Introduction

Moral guidance on how we should lead our daily lives has changed. A widespread assumption is that the welfare state, as a moral project, is old-fashioned and paternalistic. It has lost its legitimacy and more or less disappeared. Issues of responsibility are redirected from obligations toward the (nation) state and the quality of the populace to focus on the active consumer who makes conscious choices.[1] However, in this chapter the view is expressed that the role of the state in creating 'a good life' – in its social democratic version – has enjoyed widespread support and that it may be changing rather than disappearing. We may be witnessing the emergence of a new, but no less 'moral' state. The aims of this chapter are twofold: first, to demonstrate the changing forms of regulation of food consumption in Norway and, second, to show how regulatory strategies emerge as part of cultural, political, and economic processes which are highly influential, not only on the shaping of the strategies but also on their success. In spite of changing strategies, there are strong path-dependencies formed by the ways in which earlier regulatory efforts have been institutionalized.

At a time when there is a focus on individualization, globalization, creolization, and the notion of a hedonistic, choosing consumer, how should we interpret the prevailing dominance of the *matpakke*, i.e. the Norwegian packed lunch consisting of open sandwiches with hard cheese on wholegrain bread wrapped in paper?[2] And why do Norwegians eat so healthily in some respects, yet much less so in others? Introduced as a disciplinary effort, the packed lunch became integrated and institutionalized, morally and organizationally, an institutionalization which helps to explain the survival of this ascetic meal into the current period. These early regulatory efforts also had an impact when the focus was redirected in the late twentieth century towards strategies promoting self-governance. Their variable success, as exemplified by fat and sugar, demonstrates how governance is not only a strategic project; it involves and is developed within practical, organizational

---

1 Cohen 2003, Halkier 2001, Schild 2007, Sulkunen, Rantala and Määttä 2004, Neumann and Sending 2003.
2 Bauman 2001, Schulze 1992, Sulkunen 1997.

and normative frames which contribute to stability and uniformity of eating, but also produce tensions and difference.

The empirical point of departure is a series of observations about Norwegian food culture made through cross-country comparisons as well as in-depth national studies. One comparative study is about the daily patterns of eating in the Nordic countries.[3] Another deals with social and institutional conditions for consumer trust in food in Europe.[4] A discourse analysis has explored the context of Norwegian meals;[5] while a fourth approach is a structural study of food in Norway.[6] A number of studies of nutrition and food policy have also been drawn on. These indicate some quite distinctive features of the Norwegians' relationship to food and food consumption. In a comparative perspective, Norwegian eating habits are characterized by homogeneity, simplicity, and strong references to norms about healthy and proper eating, with a strict division of everyday and leisure spheres. Moral puritan values are intrinsic to everyday eating, as reflected in modest meals. Compared with the other Nordic countries, Norwegian meals generally include fewer items, with a limited range of dishes.[7] Eating associated with leisure is rather the opposite. Norwegians' consumption of soft drinks and ice cream is among the highest in Europe. Although Norwegian eating patterns are changing and differentiated, cross-country comparisons indicate that the perhaps over-simplified description presented here has some persistent and valid features.

The History of the Packed Lunch

Whereas food habits generally change slowly, more in terms of calibrations than ruptures, the *matpakke* is an example of a complete culinary revolution. The *smørbrød*, or the open sandwich as a specific dish, seems to have arrived in Norway from Denmark and Germany in the nineteenth century, perhaps introduced by civil servants, many of whom had strong connections to Denmark.

However, the status of the packed lunch cannot be understood merely as a diffusion of North European culinary traditions or as a result of industrialization. It is an institution invented by hygiene and nutrition scientists, first presented as rules about healthy school meals. With the discovery of vitamins and other micronutrients during the 1920s and 1930s, attention was directed towards the 'quality' of food and the composition of the diet.[8] In the 1920s, a leading

---

3   Kjærnes 2001.
4   Kjærnes, Harvey and Warde 2007.
5   Bugge and Døving 2000.
6   Døving 2003.
7   Mäkelä 2001.
8   Lyngø 2006.

hygienist and the head of municipal school health services, Professor Carl Schiøtz, formulated a plan to replace inferior cooked food.[9]

But during the interwar years, Norwegian municipalities were very poor, Oslo being a rare exception. With sharp conflicts over social benefits for the large numbers of unemployed, most municipalities could not afford costly reforms like the introduction of school meals. One solution was 'the Sigdal breakfast', named after the community where a country doctor, O. Lien, introduced his own version. Lien, who was part of the hygiene movement, wrote the following in a Norwegian medical journal:

> The Oslo breakfast was, on the initiative of Professor Schiøtz, introduced in some schools in Oslo in 1932...but in the countryside...such an arrangement would be unfeasible...Everybody agrees that an improvement of the diet in rural communities is imperative. The question is the method...An Oslo breakfast brought along (from home) is easy to realize without extra financial contributions.[10]

Eureka! The end result of this initiative was that within a few years two meals in Norwegian food culture changed, producing a culinary divergence among Nordic cultures that still exists seventy years later.

Political Consensus on National Welfare

Another issue influenced the regulation of eating, namely support for Norwegian agriculture, particularly the dairy industry, which became increasingly important during the interwar period. Nationalistic references to the Norwegian farmer and domestic produce, combined with socialist solidarity with poor smallholders and successful interest politics by farmers' organizations, produced a consensual understanding of the need to support dairy farmers financially as well as by drinking milk.[11] The Labour Party took over the government in 1935 based on a political agreement with the Agrarian Party. This represented a milestone. Deep-seated Labour Party scepticism towards farmers and their claims for higher food prices was now replaced by the influential slogan 'city and land – hand in hand'. Attention shifted from urban to rural poverty. Feeding the nation – a nutrition plan – was a key element of this new support to agriculture.[12] Improving nutrition through increased milk consumption could, at the same time, yield higher incomes for the dairy farmers. The potential conflict over food prices was to be resolved through better income distribution, higher social benefits, and careful price regulation. The

---

9   Døving 2003, Lyngø 1998.
10   Lien 1936.
11   Kjærnes, ICREFH III, 1995.
12   Jensen, ICREFH II, 1994.

outcome was that the National Nutrition Council developed what a contemporary conservative professor of nutrition, Ragnar Nicolaysen, ironically characterized as 'milk propaganda'.[13] The new school meal concept, recommending half a litre of full-fat milk daily for each child, fitted very well into this new policy.[14] Adults were advised to drink a full litre.

With very limited public funding, the school meal of a packed lunch united several good causes that later promoted the development of a proud identity for those who appreciated the rhetorical trope of 'a young nation': improved nutrition for the poor as a public responsibility, puritan ideals of simple, uncooked food for personal and national discipline, and support for domestic farmers. The solution was also in accordance with the ideology of leading economists and the labour movement, which insisted that improvement of social welfare should take place by seeking to raise incomes, unlike the Swedish focus on extending public services.[15] Hygienist social reformers and social democrats shared the view that meals should remain a private responsibility, to be regulated only by (paternalistic) education and, according to the social democrats, by ensuring sufficient resources for everybody.[16]

While these ideas were developed before World War II, they were realized on a national scale mainly during and after the war, at a time when the improvement of welfare via higher domestic food production and collective efforts and sacrifices by all citizens was a key issue in Norwegian politics. For example, the importation of fruits was strongly restricted and the advice was to replace them with raw carrots or – preferably – raw yellow turnip, which contains more vitamin C and is easily grown in Norway.[17]

Governing Meals

Gradually, and first among urban employees and workers, open sandwiches brought from home replaced the cooked midday meal. The disciplinary project had enough sway, logistics and legitimacy to extend the *matpakke* lunch beyond the school context. Today, whole wheat bread spread with margarine and thinly topped with cheese, brown whey cheese, salami, or liver pâté completely dominates the Norwegian lunch menu.[18] And people like it.[19] Most of those eating away from home bring a packed lunch wrapped in sandwich paper. This was truly a top-down

---

13   Kjærnes 1990.

14   Lyngø 2006.

15   Befolkningskommisjonen 1938, Wold 1941.

16   Jensen and Kjærnes 1997.

17   The yellow turnip is also known by the names 'swede' (United Kingdom) and 'rutabaga' (America).

18   Kjærnes 2001.

19   Bugge 2007.

reform, where the main concern was health and nutritional status and the national good, especially for the producers, not public demand.

Little is known about whether people initially liked the packed lunch with its milk and raw yellow turnip slices. People were not asked or monitored as they are today. But we do know that after interwar poverty and wartime hardship the average culinary standard was low. There seems to have been widespread confidence in the modernist and positivist project of 'a rational diet'. National dietary exhibitions were a public success.[20] As with all successful projects regulating action, the packed lunch became the norm. It is, however, questionable to what degree this implies self-governance. The *matpakke* institution represents very strict rules and leaves little to personal flexibility.

The means of change were the children, who became both the subject and the object of regulation. A crucial step was the transformation of the 'Oslo breakfast' provided in schools into a strongly monitored packed lunch. At the end of his famous article that launched the *matpakke*, Lien wrote:

> Finally, I would like to mention another advantage of this dietary arrangement, that when the schools are inspected, the homes are as well. Every home has to obtain this food for the children, thus the same food is to be found at home. With a school meal the child may get the 'Oslo breakfast' at school, while at home they may perhaps go on with their former habits. There is reason to believe that if the 'Oslo breakfast' brought from home can be introduced, it will affect the popular diet more deeply than a meal served in school.[21]

Here was a movement with strong beliefs and enthusiastic participants that set up a regulatory strategy within a context of political consensus. They had both the dedication and the power to prescribe. Here Lien had made a prophecy that came true. This was achieved and accomplished: institutionally centralized policies managed through disciplinary actions that were explicitly sanctioned. The monitoring of the homes was made explicit through the children. Eventually, however, the official policy with direct monitoring of what the children were eating was converted to individual and cultural sanctions in the forms of righteous indignation, guilt, and shame. This combination has proved to be particularly potent. Seventy years later, parent-teacher meetings are filled with the discourse of damning packed lunches that include pizza or noodles or chocolate spread on children's sandwiches. Arguments used refer only to health, not to a moral or political discourse.

The *matpakke* became a national icon. A discourse analysis of the *matpakke* seventy years after its invention reveals that eating a *matpakke* is 'being Norwegian', synonymous with being strong, disciplined, slim, tall, with good teeth and digestion, and high moral values, as opposed to fat Americans, mannered

---

20 Lyngø 2006.
21 Lien 1936: 547.

French, or state-indoctrinated Swedes. Discarding the *matpakke* is synonymous with a possible degeneration of values that 'built the country', culture and nation. Its distinctive normative framing may be traced back to the combination, and harmonization, of norms and structures on the family side with historical roots in the Protestant work ethic, with the needs of the agricultural market and collective political goals. Some results are paradoxical or surprising. First, the privacy of eating: eating should take place at home, or in other people's homes. If eating takes place away from home, as when at work, in school, or on a Sunday trip, food should be brought from home in the form of a packed lunch. Then food from home must somehow be eaten in public. Second, puritan and egalitarian values persist within a setting of affluence and increasing social differentiation: everyday meals and cooking should be simple and plain and not 'luxurious' in any respect.[22] These processes have produced consensus and legitimacy as well as an organizational structure that may explain why the packed lunch has survived so remarkably well. Habits embedded in practical organizational solutions were strongly supported and offered few alternatives. This was combined with the internalization of strong norms of discipline within frames of collective welfare and equality.

The Institutionalization of Food Consumption: Everyday Modesty and Leisure Hedonism

With the Norwegian industrial revolution of the late nineteenth century, a distinction was made between working time and leisure. These two spheres were soon morally divided into ascetic working time and hedonistic leisure.[23] In the work ethos, the postponement of needs is crucial: you can't have your pudding if you don't eat your meal. In the same way, leisure is a reward for work accomplished, and people without employment do not have leisure time. This division between the spheres of work and leisure is very dominant and consensual in Norway and it links directly to norms about eating. People who break the rules of the work ethos, for example by drinking alcohol or consuming sweets within an everyday context, are sanctioned by everything from subtle disapproval to official intervention in cases of deficits in parental care. The *matpakke* is clearly part of the work ethos, a kind of embodiment of this ethos. The reverse also applies; by not enjoying leisure time, you break the rule of a harmonious Norwegian ethos. Most foodstuffs and alcohol are classified in terms of self-discipline versus indulgence. Sweets are supposed to be consumed at the weekend according to the institution *lørdagsgodt* ('Saturday sweets'), as a treat after work and proper dinners during the rest of the week. On special occasions sweets are not only allowed but even prescribed. For many Norwegians, Christmas celebration and holidays without sweets are unheard of.

---

22   Døving 2003.
23   Døving 2003.

In this structural division of time, the leisure sphere is private and more individualized. The norms of leisure represent pleasure as well as freedom from the rules of eating that advocate self-discipline and from politics. There are few means for politically interfering in this private sphere of leisure when people are expected to 'relax' from societal obligations and rules. You cannot bring chocolate to school, but the health visitor cannot come to your home to inspect it at weekends. Since food is so strongly normatively regulated to be ascetic and healthy, sweets, cakes and soft drinks may not even be categorized as 'food' but as a treat or a socially acceptable vice.[24] Comparatively speaking, Norwegian meals are modest and ascetic, yet consumption levels are high for chocolate, sweetened soft drinks, salty snacks, ice cream, and fruit, all of them items consumed outside regulated meals.[25] While such items add considerably to Norwegians' intake of sugar, they are grossly underreported in dietary surveys. Thus, sugar is part of the private leisure sphere and difficult to control. Fat, on the other hand, has usually been associated with milk, butter or margarine, cheese, and meats, or in other words, food items that are part of the official 'proper' meals: the *matpakke* and the equally important family dinner.

Fat and Sugar: Different Scope for Societal Action

Nutritional problems were redefined in the 1960s and 1970s, at a time when the traditional disciplinary approach of monitoring one meal was no longer applicable and new, more individual and persuasive strategies emerged to make people change their habits. However, leaving flexibility and responsibility to the individual was not at all straightforward, and the new policies did not represent a clean break with past practices. On the contrary, the established policies had obvious implications for how the new problems were handled. The contrasting cases of fat and sugar show how this path-dependency is strongly linked to the ways in which food consumption and food policies were institutionalized.

Well into the twentieth century, fat was regarded as an efficient source of energy and beneficial for health. A diet rich in fat was seen as an important remedy against the much-dreaded tuberculosis. Cherished recipes included large amounts of cream, butter and eggs.[26] This positive perception of fat of animal origin is evident in the design of the reformed school meal.[27] Butter and milk, produced by an increasingly well-managed industry, cooperatively owned by the dairy farmers, became key elements of the new food policy based on subsidies and import tariffs.[28] Strong economic interests were involved in the production

---

24   Døving 2003.
25   Kjærnes 2001.
26   Notaker 2001.
27   Haavet 1996.
28   Kjærnes, ICREFH III, 1995.

of margarine as well, it being the main end-product from herring fisheries and the whaling industry.[29] In this competition between domestic interests, political interventions systematically favoured dairy fat. However, from the interwar years onwards, increased buying power and general market expansion meant that both industries earned good money for several decades. In this way fat was linked to important domestic interests and extensive political intervention.

Health problems associated with high intakes of saturated fat were already recognized by Norwegian nutrition experts in the late 1950s and early 1960s. Dairy fat, meat and margarine were redefined from good to bad. The most significant and politically-supported agricultural sectors were affected, and the welfare state had produced a problem for the wellbeing of the population. This meant a shift in the nutritional status of whole milk from its position as *the* good food to its new role as a major nutritional problem.[30] With its particular significance, symbolically, politically and economically, the struggle over dairy fat became a crucial issue.[31] This was a challenge for policy bodies like the National Nutrition Council, which based its recommendations on consensual, negotiated solutions involving all concerned parties.

However − and this is the surprising observation − the welfare state had produced institutions that formed a basis, a point of departure, for reformulating and reorganizing regulatory efforts to address the new problems. When discipline was no longer seen as feasible, what was then to be done? Here we see the strong path-dependency of Norwegian nutrition policy. Dominant beliefs in political solutions led to a reinforced emphasis on 'structural' or 'indirect' measures, addressing the vast corporate system of agricultural planning and subsidies, plus state-led negotiations with agricultural organizations.[32] One might critically say that the reformulated policy is an example of adaptation, that is a social problem being reformulated so that it suits the cause of powerful interests. And the new regulatory efforts were in fact rather defensive. But new dietary goals were legitimized and the promotion of animal fat was gradually phased out, i.e. a kind of de-regulation. Nutrition having been established as a public and political responsibility rather than a private and individual one, ordinary people as eaters and buyers were accorded neither influence nor responsibility. The underlying tensions are well illustrated by the case of low-fat milk. Due to resistance from the dairy monopoly, it took fifteen years of entreaties by nutrition experts and consumer groups before a low-fat variety of milk was introduced on the Norwegian market in 1984. After its introduction it took just a couple of years for the majority of Norwegians to switch from whole milk to this new alternative, accompanied by a drop in the fat content of the average diet.[33] All in all, the 'national menu' was revised, but not

---

29   Fjær 1990.
30   Kjærnes, ICREFH III, 1995.
31   Fjær 1994.
32   LD 1975.
33   Kjærnes, ICREFH III, 1995.

dismantled. Fat did re-politicize food, but the critique was soon recaptured by a producer-dominated agenda along established alignments of the welfare state: welfare for people *and* for the farmers.

Sugar, on the other hand, was never part of the Oslo breakfast or of any gastronomic discourse. Sugar was the dentists' enemy number one. Nutritionally, sugar was just 'empty calories'. Unlike in Britain, for example, sugar played a marginal economic role in Norway.[34] Sugar or syrup as such was not produced domestically but sourced globally, and most of it was sold at a low price without further processing. It has had no position in the welfare state. Market regulation was limited to a 'luxury tax' imposed on sweets and chocolate in 1933, to be reinstated several times but dismantled in 2001.[35] Chocolate and soft drink manufacturers have been concerned with freedom from state intervention, especially taxation, rather than protection. They have concentrated on competition and the marketing of products for individualistic and pleasure seeking 'new consumers'.[36] The chocolate industry thrived. Entrepreneurial chocolate manufacturers introduced new technologies and innovative marketing techniques, some of which, inspired by British and American chocolate manufacturers, included references to social and nutritional consciousness. While ingredients like cocoa and sugar were imported, it was the milk content that was emphasized in national advertising campaigns, and the interests and legitimacy of chocolate manufacturers do not seem strongly associated with sugar. Even as new health problems related to sugar were recognized, politically sugar remained a minor issue.

But parallel to this, and with increasing force in the 1980s, a new policy developed in health education that combined nutrition with other behavioural or lifestyle issues like smoking and exercise.[37] Through the use of education and campaigns, people were to be informed and persuaded take responsibility for their own health as formed by their personal way of life. Health education addressing fat, which often included specific advice on how to change the contents of major meals, was added to the politically and structurally oriented policy. Sugar, a minor part of ordinary meals, was addressed less as a dietary reform, often decoupled as 'sugar behaviour', or a case for market regulation; its regulation relied solely on arguments about the unhealthiness of sweet food items. This difference was not accidental; it is strongly linked to the ways in which these two food components were institutionalized.

The market indirectly has become more and more an arena for promoting self-governance. The sugar market has been extended and competition enhanced by the introduction of 'light' products, i.e. products where artificial sweeteners have replaced sugar. Among meat producers, generally represented by near monopolies which compete with each other, the white meat sector actively promotes its low

---

34   Mintz 1985.
35   Døving 2002.
36   Rudeng 1989.
37   SD 1992.

fat content as a competitive advantage. For the monopolistic dairy industry, it is hard to know whether the introduction of low-fat products is an effect of market strategies or public policy-making.

Between Social Discipline and Self-Discipline

The *matpakke* represents a powerful disciplinary reform fitted into the wider political project of building the welfare state. But from the invention of the packed lunch in 1936 until today, Norway has changed from being a poor country into a wealthy social democratic state with more individualistic people. We see the survival of the *matpakke* institution as an indication that there has not been a corresponding shift towards a more liberal diet in terms of individualization or a reduced significance of shared norms. The ascetic meal system has been maintained, a structure that holds work as the source of rights. This does not mean that no changes are taking place. But liberalization seems to be 'seeping in' gradually and at the margins of this meal system.

The case of the packed lunch shows how Norwegian food and nutrition policy has played an active part in shaping Norwegian food culture and food supply. The impact on the current regulation of dietary habits has been analysed using two particular cases, fat and sugar. Encouraging a reduction in the consumption of animal fat has to a large degree succeeded, while advice regarding sugar is noticed, but less often put into practice. Table 10.1 summarizes the distinctive features of the societal handling of excess fat and sugar, respectively. A number of distinctive features can be identified.

Table 10.1   The Norwegian Approach to Nutritional Problems of Fat and Sugar

	Fat in Excess	**Sugar in Excess**
Production and market	Basic element of domestic agricultural production (esp. dairy)	Imports
	Long-standing conflict with other domestic fat producers in a saturated market (margarine)	Little competition in an expanding market
	Long history of market regulation, including price	Almost no market regulation, low prices (except during a period with a 'luxury tax')

	Basic component of meals	Not part of meals, but consumed as individual treats and on special occasions
Consumption and households	Emerged as a potentially lethal problem for middle-aged men – a general welfare concern	Emerged as a dental problem for children – not a general welfare concern
Nutrition policy	Links to established political conflicts	No established political conflict alignments (e.g. consumers v. producers)
	A social and political problem – an integral part of nutrition policy (redefined from beneficial to detrimental)	A private problem of personal discipline – until recently external to nutrition policy

One major difference is related to their respective normative framing within the Norwegian system of eating. The explicit regulation of eating has been and still is strongly associated with proper meals, including both sandwiches and cooked meals, but opposed to snacks. Meals of a certain structure are normal; snacks represent a tempting sin and are perhaps not regarded as food at all. The disciplined everyday meals have traditionally contained considerable amounts of fat, but little sugar, which was, and is, mostly relegated to the leisure sphere. Reducing fat intake was about reforming *social* meals, while reducing sugar intake involves requests for self-discipline in terms of limiting *individual* snacking. Norwegians are motivated for self-discipline in workday meals but not when enjoying the time off that they have earned. The norms of leisure are the opposite of the work ethic, allowing for hedonism and indulgence. In this private sphere of freedom, arguments of self-discipline are largely irrelevant.

Second, there are major differences in how fat and sugar markets are structured and regulated. These differences, too, are associated with the development of the Norwegian welfare state. Fat was, and is, associated with the dominant agricultural industries and extensive government intervention through the public sphere, while sugar is only marginal in all these respects. Change might be expected to take place more easily within an area less dominated by strong opposing interests and habits. The problems of fat, being part of the collective building of welfare, were taken seriously as social and political problems. For sugar, relevant policy-making forums were few and far between. As a private problem, sugar is outside the reach of the established regulatory measures, while fat could be governed by economic and normative means within the system.

Third, compared to the major issues of agriculture and welfare, the political story of sugar is weaker. Even though there are Norwegian bad guys ('sugar pushers') in this story, these are rarely visible in public discourse. Mobilization has mainly focussed on multinational corporations, with few means for politicizing sugar in the Norwegian context. Hardly any Norwegian is unaware of the advice to cut down on sugar. Yet the degree of success is meagre.

There is an opposition between the institutionalization of fat in the diet as a public and social concern and sugar as a private and individual matter. There is a dynamic where the regulation of food consumption has powerful effects on what people eat when there is accordance between regulations, market institutions, and cultural and normative references for everyday habits, and much less – or in unexpected directions – when this is not the case. Harmonization of different social institutions has been particularly strong in Norway because of specific interrelations between households, the food market, and regulatory institutions within the welfare state. This configuration has been capable of incorporating even new and redefined goals when they fit into the established institutional field. The connection between self-responsibility and collective welfare, fortified by the institution of the packed lunch, helps to explain the very rapid and wide-reaching switch from full-fat to low-fat milk as soon as it was made available. Outside that institutional field Norwegians can and do exercise their individual freedom in the market and in the family – for example by enjoying large amounts of sweets, soft drinks, and ice cream. What may be questioned is therefore the degree to which we are observing a substantial shift can be observed towards self-regulation and self-responsibility with regard to nutrition.

These findings are important for dietary policies to combat obesity. First, the social constructions underlying the handling of fat in public policy and everyday practice have had effects on consumption levels. These may, in turn, be one reason why Norwegians are comparatively speaking less obese. Second, the collective and universal framing of habits as well as policies may point in the same direction. Third, however, the incidence of obesity is growing in Norway as well. The findings in this paper point to the taken-for-granted institutional and normative structures which form the foundations for what we do in politics and in everyday life. Policies need to acknowledge such historically contingent structures, to align with or counteract them. The case of sugar indicates that this is no minor task.

References

Bauman, Z. 'Consuming Life', *Journal of Consumer Culture* 1, 2001, 9–29.

Befolkningskommissionen *Betänkande i näringsfrågan. Sveriges Offentliga Utredningar 1938: 6* [Report on the Food Issue. Sweden's Public Investigations] Stockholm, 1938.

Bugge, A. *Ungdoms Skolematvaner: Refleksjon, Reaksjon eller Interaksjon* [School Meals among Young People: Reflection, Reaction or Interaction], *SIFO Report No. 4: 2007*, Oslo, 2007.

Bugge, A. and Døving, R. *Det Norske Måltidsmønsteret – Ideal og Praksis* [The Nordic Meal Pattern – Ideal and Practice], *SIFO Report No. 2: 2000,* Oslo, 2000.

Cohen, L. *A Consumers' Republic: the Politics of Mass Consumption in Postwar America*, New York, 2003.

Døving, R. *Mat som Totalt Sosialt Fenomen. Noen Eksempler med Utgangspunkt i Torsvik* [Food as Total Social Phenomenon. Some Examples Based on Torsvik] Unpublished PhD Dissertation, Department of Social Anthropology, University of Bergen, 2002.

Døving, R. *Rype med Lettøl. En Antropologi fra Norge* [Grouse with Beer: an Anthropology from Norway], Oslo, 2003.

Evang, K. and Hansen, O.G. *Norsk Kosthold i Små Hjem* [Norwegian Diet in the Poor Households], Oslo, 1937.

Fjær, S. *Makt, Marked og Margarin. Fettreguleringens Politikk* [Power, the Market and Margarine], *SIFO Report No. 6: 1990,* Oslo, 1990.

Fjær, S. 'Forhandlinger, Fettkabaler og Forebygging' [Fat: Negotiations, Interests and Prevention], in Elvbakken, K.T., Fjær, S. and Jensen, T.Ø. (eds), *Mellom Påbud og Påvirkning. Tradisjoner, Institusjoner og Politikk i Forebyggende Helsearbeid,* [Traditions, Institutions and Politics in Preventative Health Care] Oslo, 1994, 149–58.

Halkier, B. 'Consuming Ambivalences: Consumer Handling of Environmentally Related Risks in Food, *Journal of Consumer Culture* 1, 2001, 205–24.

Haavet, I.E. *Maten på Bordet. 50 år med Statens Ernæringsråd* [Food on the Table: 50 Years of the National Nutrition Council], Oslo, 1996.

Jensen, T.Ø. 'The Political History of Norwegian Nutrition Policy', in ICREFH II, 1994, 90–112.

Jensen, T.Ø. and Kjærnes, U. 'Designing the Good Life: Nutrition and Social Democracy in Norway', in Sulkunen, P., Holmwood, J., Radner, H. and Schulze, G. (eds), *Constructing the New Consumer Society*, Basingstoke, 1997, 218–33.

Jones, G. *Social Hygiene in Twentieth Century Britain*, London, 1986.

Kjærnes, U. *Velferdskrav og Landbrukspolitikk. Om Framveksten av Norsk Ernæringspolitikk* [Welfare Needs and Agricultural Policy: the Emergence of Norwegian Nutrition Policy], *SIFO Report No. 7: 1990*, Oslo, 1990.

Kjærnes, U. 'Milk: Nutritional Science and Agricultural Development in Norway 1890–1990', in ICREFH III, 103–16.

Kjærnes, U. *Eating Patterns. A Day in the Lives of Nordic Peoples. SIFO Report No. 7: 2001*. Oslo, 2001.

Kjærnes, U., Harvey, M. and Warde, A. *Trust in Food. A Comparative and Institutional Analysis*, Basingstoke, 2007.

LD *Om norsk matforsynings- og ernæringspolitikk* [Norwegian Food Security and Nutrition Policy]. *Ministry of Agriculture. Stortingsmelding no. 32 (1975–76)*. Oslo, 1975.

Lien, O.L. 'Noen Betraktninger over Skolebarns Ernæring. Hvorledes en Kostreform kan Innføres i Skolene Også på Landsbygden' [Some Reflections on School Children's Nutrition. How a Dietary Reform can be Introduced in

Schools in Rural Areas], *Tidsskrift for Den Norske Lægeforening* 56, 1936, 540–47.

Lukes, S. *Power: A Radical View*, Basingstoke, 2nd ed., 2005.

Lyngø, I.J. 'The Oslo Breakfast – an Optimal Diet in One Meal. On the Scientification of Everyday Life as Exemplified by Food', *Ethnologica Scandinavica* 28, 1998.

Lyngø, I.J. *Vitaminer! Kultur og Vitenskap i Mellomkrigstidens Kostholdspropaganda,* [Vitamins! Culture and Science in the Inter-war Propaganda on Diet], Unpublished Dr Art Thesis, University of Oslo, 2006.

Mintz, S.W. *Sweetness and Power: The Place of Sugar in Modern History*, New York, 1985.

Neumann, I.B. and Sending, O.J. *Regjering i Norge* [Government in Norway], Oslo, 2003.

Notaker, H. 'Nasjonal matkultur' [National Food Cultures], *Nytt Norsk Tidsskrift* 2001, 205–6.

Rudeng, E. *Sjokoladekongen* [The King of Chocolate], Oslo, 1989.

Schild, V. 'Empowering "Consumer-Citizens" or Governing Poor Female Subjects?: The Institutionalization of "Self-Development" in the Chilean Social Policy Field,' *Journal of Consumer Culture* 7, 2007, 179–203.

Schiøtz, C. *Steller Norge Forsvarlig med sine Skolebarn og sin Ungdom? Nei!* [Does Norway Deal with its Schools and its Children Properly? No!], Kristiania, 1917.

Schulze, G. *Erlebnisgesellschaft. Kultursoziologie der Gegenwart* [Experiential Society. Sociology of Contemporary Culture], Frankfurt, New York, 1992.

SD *Utfordringer i Helsefremmende og Forebyggende Arbeid* [Challenges in Health Promotion and Prevention Work], *Ministry of Social Affairs. Stortingsmelding no. 37 (1992–93)*. Oslo, 1992.

Seip, A.L. 'Fattiglov og Fattigvesen i Mellomkrigstiden – et Forsørgelsessystem under Krise' [Poor Law and Poverty Services in the Inter-war Period. A System of Poverty Support in a Time of Crisis], *Historisk Tidsskrift* 66, 1987, 276–300.

Sulkunen, P. 'Introduction: The New Consumer Society – Rethinking the Social Bond,' in Sulkunen, P., Holmwood, J., Radner, H. and Schulze, G. (eds), *Constructing the New Consumer Society*, Basingstoke, 1997, 1–20.

Sulkunen, P., Rantala, K. and Määttä, M. 'The Ethics of Not Taking a Stand: Dilemmas of Drug and Alcohol Prevention in a Consumer Society – a Case Study,' *International Journal of Drug Policy* 15, 2004, 427–34.

Wold, K.G. *Kosthold og Levestandard. En Økonomisk Undersøkelse* [Diet and Living Standards: an Economic Survey], Oslo, 1941.

PART 3
Social and Medical Influences

Chapter 11

Diet, Body Types, Inequality and Gender: Discourses on 'Proper Nutrition' in German Magazines and Newspapers (c.1930–2000)

Jürgen Schmidt

Introduction

To talk about 'proper nutrition' and body types seems in some basic respects to be ahistorical with regard to the twentieth century. To begin with, the concept of the ideal human body was fairly stable. The ideal body was slim and strong, not stout or flabby. During the nineteenth century some degree of corpulence had signalled success and wealth, but this view was less acceptable in the twentieth century. Social distinction was no longer expressed by a corpulent body but by the what, where, and how of eating. Despite all the efforts of experts, institutions, state bureaucracy and the media, every generation suffers from new types of nutritional maladies. In recent times, obesity, anorexia, and bulimia have all captured the attention of the public.

This chapter will seek to answer four questions: first, who defines what proper nutrition is? Second, how did recommendations for proper nutrition develop? Third, how and to what extent were the recommendations realized, and fourth, what was the relationship between proper nutrition and twentieth-century body ideals?

Proper nutrition cannot be defined in advance because the meaning of what 'proper' was and is always depends on the historical context and this changed during the course of the century. Here two things should be borne in mind. First, proper nutrition in Germany could not be analysed in wartime or in the immediate post-war period. Proper nutrition cannot be defined in that narrow sense to find the answer to the question of how to feed the inhabitants of a country or a city in times of scarcity.[1] Proper nutrition means that there is enough food so that the majority of people will have the possibility of choosing how to regulate their eating behaviour in the light of nutritional recommendations.[2] Secondly, what becomes obvious, too, is the fact that the two extremes of human physique – thin and obese – were both seen as 'improper'.

---

1   See for this aspect Schmidt 2007.
2   See for this approach Barlösius 1999: 220.

To research these questions an analysis has been made of a variety of magazines, newspapers, and websites – for women (housewives) and for men – from the 1920s to the present day.[3] My main focus is on popular media, and excludes journals written for professional nutritionists. However, this research reflects two different assumptions. On the one hand, there is the supposition that body types are constructed by discourses – the norms, pictures, and texts that influence our perception of our body. On the other hand, it can be said that body types are formed by nature – size, weight, sex, age, and nutrition are all very real factors which determine the body form.[4]

For Arthur Scheunert, a Professor in Leipzig and one of the founding fathers of vitamin research in Germany, defining proper nutrition was very simple. He proclaimed in '*Das Magazin der Hausfrau*' (the Housewives' Magazine), a popular magazine with a regular circulation of 237,500 copies:

> Nutrition is only proper when all substances that we need for living are administered with our daily meals and, at the same time, enough fuel is provided for our body to perform its daily tasks.[5]

The decisive elements for Scheunert were fat, carbohydrates, protein, mineral nutrients, vitamins, and water. In 1954 E. Vetter defined proper nutrition in nearly the same way in an article on 'The basic law of the menu' in the magazine *Wir Frauen im ländlichen Leben* (Women in rural life):

> Our food has to be wholesome. It must contain all the elements that our body needs to live, meaning that all life processes should proceed as undisturbed as possible by diseases so that it can work, e.g. that we can do our full daily work without particular exhaustion.[6]

Finally, in 2006, in a long essay in the weekly newspaper *Die Zeit* it was stated that:

> It is clear that nutrition has to cover the vital needs of our body. Without protein, some kinds of fat, vitamins, mineral nutrients, trace elements and an adequate energy supply, our body will perish.[7]

---

3   *Das Magazin der Hausfrau* 1928–1929, *Hannoversche Land- und Forstwirtschaftliche Zeitung* and supplement *Wir Frauen im ländlichen Leben* 1954–1956, 1962, *Wir Hausfrauen. Mitteilungsblatt des Verbandes Berliner Hausfrauen* 1956–1958, 1962–1963, *Brigitte* and *Men's Health*, web edition 2006.

4   See Lorenz 2000: 94–6, for a critique of Foucault's one-sided view of the body only in the sense of a discourse, ignoring the 'physical-psychic entity'.

5   *Das Magazin der Hausfrau*, Issue 74, 21 October 1928.

6   *Wir Frauen im ländlichen Leben*, No. 5, 30 October 1954: 1325.

7   Herden 2006.

This is the 'general line' which Uwe Spiekermann criticized in an article about nutrition objectives as structural conservatism in German nutrition research. Natural science gives a formula – 'enough' and 'the right mixture' – to which people as 'natural beings' had to adjust. Factors beyond the 'material-physiological' side of nutrition were ignored.[8]

Sabine Merta emphasizes that at the beginning of the twentieth century, corpulence, especially in rural areas, was still associated with positive adjectives such as 'sturdily built' or 'well fed'. Slimness as an ideal belonged to the urban middle classes. But after 1910 slimness started to feature as the ideal body form in the Western world – maybe as a late result of middle-class ideals of austerity and temperance.[9] George Hersey assumes, in contrast, that the concept of proportion in occidental art created the image of the ideal body; not only by weight itself but the whole appearance of the silhouette is decisive for what the observer regards as beautiful.[10] In the twentieth century this was the slim body for women and men; in the course of the twentieth century for men this became more than slim but also a muscular body. This prevalent ideal for women can be illustrated by a very simple example. In a magazine for women living in rural north west Germany an article gives hints on how to use stripes so that the body does not appear too fat and not too thin when making one's own clothes. In a little story 'chubby Käthe' saw 'skinny Lotte' with her new horizontally-striped dress. Käthe uses the same pattern for sewing. But the result for her was a horror: 'She was appalled. She was not only chubby, she was really fat'. Therefore she decided to use stripes which ran diagonally: 'When Käthe finished her dress she was very satisfied. With diagonal stripes she appeared much thinner than before'.[11] This story shows two things. Not all women fulfil the norms, but the slim body is the norm towards which women aspire. And today, in TV movies like '*Moppel-Ich*' ('The Tubby Self') based on a non-fiction book with the same title, the emancipated woman who is defending her right for her own body is not really fighting against the prevalent norm of slimness but accepting it.[12]

Despite nutritional recommendations and because of the norm of slimness, there have been ongoing debates about nutritional diseases. Young girls' anorexia is not only a phenomenon of our day, but has been known since the late nineteenth century.[13]

---

8   Spiekermann 2001: 98–9.

9   Merta, 2003: 286, see also Thoms 2000.

10   Hersey 1998.

11   *Wir Hausfrauen im ländlichen Leben*, No. 7, 18 February 1956: 232.

12   *Das Moppel-Ich*, Zweites Deutsches Fernsehen ZDF, 12 March 2007, Fröhlich 2004.

13   Habermas 1994, see also Chapter 15 by Ulrike Thoms.

The Structure of Recommendations for Proper Nutrition

In the present chapter nutrition recommendations mean both advice from professional nutritionists to a broader public as well as nutrition hints from journalists and editors based on professional nutritionists' recommendation. The second aspect includes the transformation of nutrition information into journalistic 'frames', especially health, fitness and, as a general topic, the male and female body.[14] The structure of nutrition recommendations in the course of the twentieth century was very similar in some aspects. Professional knowledge had to be disseminated among consumers so as to influence their eating habits. There are different expert groups; but their main similarity results from their natural science background. Their success and advancement followed from the establishment of new disciplines, like physiology, in the early twentieth century. In the Weimar Republic the *'Internationale Arbeitsgemeinschaft zum Studium der Volksernährung; Reichsverein Volksernährung* [International Working Group for the Study of the People's Nutrition], in the Federal Republic of Germany the *Deutsche Gesellschaft für Ernährung* [German Society for Nutrition] (DGE), *Auswertungs- und Informationsdienst für Ernährung, Landwirtschaft und Forsten AID* [Analysis and Information Service for Nutrition, Agriculture and Forestry] and the former *Zentralinstitut für Ernährung der Akademie der Wissenschaften der DDR* [Central Institute for Nutrition of the Academy of Science of the GDR[15]] were and are institutes defining the norms of proper nutrition.[16]

National and transnational political organizations interested in these results influenced nutrition goals, and played an important role in transmitting nutritional knowledge. For example, the figures and tables of the League of Nations on how many calories a person needed according to his or her job influenced the calculation of the food supply until the years after the Second World War. Currently, recommendations of the World Health Organization form reference points in debates about proper nutrition. Totalitarian regimes established many organizations, corporate bodies and institutions to propagate nutrition goals. The instrumentalization of nutritional research led to a direct influence on consumers' habits (for example *Verbrauchslenkung* [Consumption control] under National Socialism). The research institutes used magazines to spread knowledge to a broader audience while the media referred to these experts when they wanted to convince the readers about new eating habits.[17]

---

14   See DGE, *Ernährungsbericht* 2004: 349–50, for the 'framing approach' in regard to nutrition research.

15   This was the official name from 1972, now changed to 'The German Institute of Human Nutrition Potsdam-Rehbrücke'.

16   For the great variety of nutrition institutions in the FRG, see the webpage of the DGE. See also Barlösius 1999: 222–4.

17   For additional forms of information (leaflets, brochures, and teaching courses on cooking in schools), see also Dannenberg 2001: 77–8.

Because of the sources selected, it is not possible to make a general statement about the recipients of the recommendations. According to the magazines, women were always responsible for observing proper nutrition in the households. A column 'Hints for the young farmer' could be found in the newspaper of the Hanoverian Chamber of Agriculture but it actually only dealt with the work, health and safety, etc., on the farm and there were no articles about nutrition, household or health. These topics were reserved for the supplement *Wir Frauen im ländlichen Leben.* The strictly separate gender interests were expressed in an article in *Wir Hausfrauen* – distributed in Berlin in the 1950s and 1960s.[18] Today, nutrition recommendations in magazines and newspapers are written in a gender-neutral style and there is a great deal of information about proper nutrition for men but most non-fiction literature about proper nutrition is still addressed to women.

Finally, even the content structure of the recommendations showed similarities in the course of the twentieth century. They either contain the 'basic rules' of proper nutrition or demonstrate the importance of vital elements of our nutrition; or they celebrate some sort of food as especially healthy, rich in vitamins or dietary fibre, including few (or many) calories and condemn other foods because they are 'unhealthy'.

Changes and Continuity in the Recommendations for Proper Nutrition

Using quotation marks when referring to the words 'proper nutrition', symbolizes that it is not possible to define a generally-accepted formula.[19] Basic rules established early in the twentieth century persisted, but much change has occurred. One reason is that knowledge about nutrition developed very rapidly in the twentieth century. The model of the body as an engine and the use of kilocalories to measure energy developed in the late nineteenth century and influenced the concept of the body for the entire twentieth century.[20] This simplistic idea found its way to a broad public. In 1928 Dr. Otto Gotthilf wrote in the *Magazin der Hausfrau* that: 'The human organism to some extent represents a living oven. If someone is to do his share, he has to be fed with enough fuel – nutrition'.[21] The result of this understanding of the body and nutrition was that energy became central to the diet. 'It is necessary that the modern housewife knows what calories are', noted an author in *Das Magazin der Hausfrau* in 1928.[22]

The amount of kilocalories suggested was quite stable, too. The figures for a worker weighing 70 kg and 25 years old differ only by about 200 kcals between the

---

18 See, for example, *Wir Hausfrauen*, No. 9, September 1956: 9.
19 Barlösius 1999/2000: 51.
20 Sarasin and Tanner 1998: 26–7, 34.
21 *Magazin der Hausfrau*, Issue 40, 26 February 1928.
22 *Magazin der Hausfrau*, Issue 48, 22 April 1928. Note that the calories mentioned here are kilocalories (kcals).

late nineteenth century and the present. Carl Voit suggested 3,000 kcals in 1881, Max Rubner 3,059 in 1908 and the German Society for Nutrition recommends 2,900 kcals today.[23] The picture of the body as a machine which needs something to burn is still alive in everyday experience. 'The way to stimulate your burning of fat' was the title of an article about metabolism in *Men's Health*. 'Activate your metabolism and change from the "storage mode" into a higher gear', is a suggestion from *Brigitte* for women who want to lose weight.[24] As long as the body is perceived in this way it is inevitable that kilocalories play an important role in our reflections about eating habits. The German word *Kalorienbombe* [calorie bomb] or the suggestion not to waste your 'calorie allowance' on things you do not like to eat show the ongoing influence of kilocalories in the discourse about proper nutrition. Another trend – found especially in recent magazines and publications – points in the opposite direction. It ignores the 'dictatorship' of calorie tables, propagates the self-conscious consumer who eats what she or he likes (but not too much and not unhealthily) and demonstrates that counting kilocalories is counterproductive for dieting. 'Stop Counting Calories' was therefore the headline in *Men's Health*. But if we count how often the word 'calorie' is used in the web-articles analysed it becomes obvious that 'burning calories' remains a central issue especially when nutrition is considered within the contexts of the body as a machine for the development of a slim or muscular body – as is the case in *Men's Health*.

Talking about proper nutrition in the twentieth century always meant talking about vitamins. The history of their detection illustrates a clear connection between proper nutrition and vitamins: Vitamin deficiences could lead to serious diseases. Therefore vitamins appeared in nutrition recommendations soon after their discovery and thus reached the public media as substances which were 'lifesavers' guaranteeing health, fitness, performance, high spirits, and more.[25] Included in these positive descriptions was advice about which foodstuffs are richest in vitamins and which vitamin has which function in the body.[26]

While there is continuity in regard to body images, energy requirements, and vitamins, the assessment of different foods changed. In the case of sugar, for example, there were very different assessments in the 1920s compared with today. The retailers' magazine for housewives emerged as a strong supporter of sugar consumption. In an issue in April 1928 it reported from a lecture about German beet sugar: 'Sugar has two enemies: cane sugar abroad, malicious gossip at home. Sugar is accused of causing diabetes, bad teeth, and even more nonsense'.[27] In this case the economic interests of the retailers contradicted the recommendations of the

---

23   Spiekermann 2001: 102, table 1.

24   Websites of *Men's Health* and *Brigitte*.

25   Werner 1998.

26   See for example *Wir Frauen*, Nr. 18, 30 April 1955: 669, *Braunschweiger Zeitung*, 26 March 1994 (cited in Benterbusch 1997): 101.

27   *Das Magazin der Hausfrau*, Issue 46, 6 April 1928.

experts. However, it should be borne in mind that the 1920s were not years of plenty for many readers and they were concerned about getting adequate nourishment. Attitudes to sugar changed gradually in the 1950s and 1960s but it was a slow process. Not until the point when firms producing caffeinated soft drinks began advertising with the label 'sugar free' did a conscious change in consumption begin to occur. For this reason the actual trend of consumption of sugar in West Germany continued climbing until the middle of the 1970s, followed by a stable period until 1990, and then slowly decreasing consumption from 1990.[28]

These debates about kilocalories, vitamins, and sugar were three examples to illustrate the traditions and changes in regard to recommendations from the nutritional research point of view and their popularization in some magazines and newspapers. But it is necessary to consider changes (and continuities) of national nutrition goals in respect to state authorities and governments, too.[29] Proper nutrition in the Weimar Republic had as its main focus supplying the population with vitamins. The new knowledge about these substances and the experience of vitamin deficiency diseases during World War I led to a broad campaign for the consumption of more fruit and vegetables and a reduced consumption of meat. In a very general sense it could be argued that the new democratic political system should be based not only on a well-fed but also on a healthily fed population, where everybody should have the possibility of participating in this nutrition goal regardless of his or her social position. By propagating 'democratic' fruit and vegetables instead of 'undemocratic' meat, the state showed new ways to achieve improved nutrition. But with this strategy the national institutions for nutrition had to face two problems. Social groups which had been unable to share in the earlier high meat diet now wanted to participate in these consumer habits and demanded more (and better) meat for themselves. Fruit, vegetables and meat were foods that distinguished one social group from another. However, tropical fruits high in vitamin C were extremely expensive and most workers could not afford them.[30]

In the 1930s, under the Nazi regime, proper nutrition not only had to be healthy, it also had the function of strengthening the body and the people for the next war. The intrinsically neutral word *Vollwert* [whole foods] took on an ideological aspect under these circumstances. It also helped to establish propaganda for 'German rye' instead of 'foreign wheat' to prepare the German wholesome body for war as opposed to the substandard body.[31]

After the wartime years of scarcity, German nutrition politics faced new problems in the postwar era with growing agricultural production. 'Diseases of civilization' was the new catchphrase and it was used by the general public without

---

28   DGE, *Ernährungsbericht* 2004: 44. In the GDR sugar consumption rose steadily until the early 1990s and was higher than in the FRG.

29   See for more details, see ICREFH II.

30   Later in the GDR the lack of tropical fruits caused discontent among the population.

31   For detail about bread, see Spiekermann 2001: 103–4.

reflecting on its ideological background.[32] The recipes proposed combined earlier recommendations. The appeal to eat more fruit and vegetables remained and the word *Vollwert* again played an important role. But *Vollwert* now received more and more the connotation of 'alternative' and 'reform nutrition' and the recommendation for proper nutrition faced new scepticism: 'Is it necessary in the countryside', the magazine *Wir Frauen auf dem Lande* asked rhetorically, 'where fresh fruit and vegetables are growing right into our mouth and we don't have to scrimp on milk, butter, eggs, and meat, to speak about the *Vollwert* of our nutrition?' The magazine gave – of course – the answer 'yes' and this for three reasons. First, even in the countryside more and more refined flour (*Feinmehl*) was used; second, improper cooking destroyed the nutrients; and, third, nutrition was not adapted to different working conditions. As in Nazi times the article recommended eating whole-grain bread, and, as in Weimar times, more fruit and vegetables to get enough vitamins.[33] The same rules were propagated to the housewives in urban Berlin in the 1950s and 1960s.[34] In the following decades the discourse about proper nutrition moved more and more from 'adequate nutrition' to 'healthy nutrition'. This remains the catchphrase right down to the present day. Facing the immense health care costs and a growing population of older people, nutrition experts are again trying to set norms of proper nutrition to make the consumer's life worth living and for the state to save money in its health care system.

When the professional nutritionists meet the consumers, a third level has to be considered with regard to proper nutrition: the individual and cultural level. Nutrition recommendations face strong obstacles. Eating – even in the fast changing times of the late twentieth century – is deeply rooted in individual, familial, and regional traditions.[35] Therefore the path from knowledge to practice in this field is very long and stony. The statement 'where would mere theory be greyer than in the field of eating', made in a 1928 article about proper nutrition in *Das Magazin der Hausfrau*, is still a challenge for professional nutritionists. The knowledge about proper nutrition has indeed spread, but consumption patterns show that these facts have only partially been translated into action. For example, the *Nutrition Report*, 2004, asserted that no group of persons in Germany ate 400 g of fruit daily – the nutrition goal defined by the World Health Organization in 2003.[36]

There are several reasons for this. First, information is the instrument in modern societies for spreading knowledge. In contrast to dictatorial regimes, the

---

32   Since the late nineteenth century, right-wing publicists and professors differentiated between a higher standing 'German culture' and a less developed, modern 'Western civilization'.

33   *Wir Frauen im ländlichen Leben*, No. 3, 21 January 1956: 84.

34   *Wir Hausfrauen*, No. 7, July 1956: 3–5. Recommendations for intakes of mineral nutrients, dietary fibre, and trace elements are also provided.

35   Selter 1995: 202, see also ICREFH VIII.

36   DGE, *Ernährungsbericht* 2004: 61. The only 'pleasant exceptions' are groups of persons older than 51.

quasi-official institutes mentioned in the first section of this chapter cannot force consumption in a special direction. Information as such is a weak instrument for nutrition education. It is volatile, socially unequally distributed, and, because it is not connected with legal guidelines or even governmental sanctions, it depends on the individual consumer to be aware of it and act in his or her own interests.

Second, in addition to basic knowledge about proper nutrition there is also the problem that there is some disagreement between experts; and the more specialized nutrition research becomes the more differences occur. 'No other science is as changeable as that of eating and drinking', states an article in the weekly journal *Die Zeit* in 2006 and points to the debate about whether fat or carbohydrates are responsible for obesity.[37] And eighty years earlier, the readers of *Das Magazin der Hausfrau* were confronted with different opinions in regard to sugar, as we have seen. This leaves the consumers in a state of uncertainty, the result of which is that consumers tend to stick to their existing eating habits.

Third, nutrition recommendations have a serious enemy: advertising. For example, in 1957 there appeared in *Wir Hausfrauen* an article emphasizing in bold letters: 'The development of fat consumption in West Germany shows an alarming increase compared to other countries'; but only one month earlier the readers saw a full-page advertisement with 100 per cent liver sausages and the text gave the housewife the advice to spread their husband's bread with 'Heinz liver sausages', so that he will be able to meet 'the great demands' that daily life 'in our times involves ...'.[38] What about today, with all the advertising for slim-line products and low-fat food? The German *Nutrition Report*, 2004, comes to a sobering conclusion. Despite marketing strategies for such products the overall impression of the presentation of nutrition on television is that the foods shown in no way mirror the nutrition recommendations of the German Society for Nutrition:

> In television advertising, sweets and snacks are shown in more than a third of the spots (35.2 per cent). Often sweets can be seen in talk shows (30.1 per cent) and animated cartoons (29.7 per cent), too.[39]

And these influences, especially on children and teenagers, are not lessened by school programmes because nutrition, health or the basics of home economics are not school subjects.[40]

Finally, social and economic development and changes in our time make it difficult for consumers to apply nutrition recommendations. The flexibility and

---

37   Herden 2006.

38   *Wir Hausfrauen*, No. 12, December 1957, 4, No. 11, November 1957, third cover page.

39   DGE, *Ernährungsbericht* 2004: 361.

40   This was already proposed in the 1970s by Teuteberg and Neuloh in the official 'nutrition report' (DGE, *Ernährungsbericht* 1976: 442). The style of these recommendations was criticized by Rath 1984: 275–80.

mobility of workers, employees, and businessmen on the one hand, and high unemployment rates and reduced social benefits on the other, are all challenges and risks that facilitate improper nutrition. But consumers resisting nutrition recommendations can be seen as individual, obstinate (*eigen-sinnig* – independently minded) persons. If nutrition recommendations are understood as a norm-setting process in modern society, people's reactions can be interpreted in two ways: either they act as self-conscious, mature citizens in the sense of 'I decide for myself what is right for me' and undermine tendencies of power, control, and discipline, or as immature consumers who are unable to deal reasonably with their eating habits. Therefore nutritionists have to address their advice to the public in broad cultural and political contexts if they want to be successful.[41]

Body Ideals and Proper Nutrition

What about the body in these discourses? Nutrition recommendations – if they were to be observed – would support the ideal of the slim body. On the other hand, there is evidence of improper nutrition at both extremes: overly thin young girls and more and more overweight or even obese children. Thus it can be seen that nutrition recommendations influence – directly or indirectly – the image of the slim body. Whereas in the 1920s articles criticized the 'slim line' as a 'luxury version' of the body which was very unhealthy,[42] there are no such interpretations today. Nowadays, 'attacks' concentrate on extremely skinny models as bad examples for girls. But in this context it is not 'proper' nutrition that is at the centre of interest – only 'improper' nutrition. The effect of actions like the ban of thin models from the catwalk seems to be very limited: 'Eating disorder activists said many Spanish model agencies and designers oppose the ban and they had doubts whether the new rules would be followed'.[43]

At present, magazines for both men and women construct clear images of the 'right body'. For women the slim body is the reference point. This is the result of a long tradition: 'Food and figure tolerance were narrower for women than for men'.[44] For men the ideal body is muscular: abdominal muscles especially are of particular interest. From ancient statues and the history of weight training and bodybuilding, a certain tradition for this kind of male silhouette can be identified but it is less pronounced than the body ideal for women.[45] The nutrition recommendations in this respect always combine nutritional advice with sports and gymnastics. The main goal is not proper nutrition, but the perfect body; nutrition

41   Spiekermann 2001: 107, Tanner 1996: 402, Barlösius 2001: 115–6, 123–4.

42   *Das Magazin der Hausfrau*, Issue 83, 16 December 1928.

43   Available at: http://www.cnn.com/2006/WORLD/europe/09/13/spain.models [accessed: 13 September 2006].

44   Schwartz 1986: 210.

45   See for the British example: Zweininger-Bargielowska 2006: 595–610.

in these articles is subordinate to the body ideal. This is the 'frame' within which journalists and editors give nutritional advice. Until the 1960s, the form of the body and its silhouette was subordinate to proper nutrition. With new liberties and the very visible process of individualization since the late 1960s, the use of this metaphor has changed. Now it cannot be found in nutrition recommendations but it is used in connection with fitness and weight training to bring the individual body into shape.

Therefore, in general, it can be seen that discourses about proper nutrition and body ideals are linked, but not as obviously as one would think at first glance. They are interrelated but with different focal points. If the consumer follows the 'right-nutrition' recommendations, she or he may get a slim body, but this is not the main purpose of the recommendations. If the consumer follows the advice for a slim body she or he may eat 'right', even though that was not the main aim of the suggestions.

Conclusion

During the entire twentieth century, in general, the discourse on proper nutrition was not simply a fight against malnutrition and obesity. Instead, it was a norm-setting process to adapt body types to the societal requirements of the times. Either the strong body fit for military services was propagated or the healthy body that caused minimum health expenditure. But these national nutrition goals faced obstacles. Cultural traditions (for example regional and familial eating habits), and individual preferences and tastes (such as meat instead of vegetables), dominated body ideals (e.g. thin female models) and restricted their influences. Even in times of massive consumption control, like the years between 1933 and 1945, there was only limited success. Today, in liberal times, recommendations for proper nutrition are based on information. This weak instrument will only be successful if the consumers are targeted in their cultural and social environment and if they understand and accept that state institutions and nutrition experts are interested in their well-being and not primarily in the requirements of the modern state. Therefore, the long process from undernutrition to obesity in the twentieth century is as much an indication of the limits of nutrition recommendation as a hint that the problem of obesity today cannot be solved by suggestions from state initiatives alone.

References

Barlösius, E. *Soziologie des Essens. Eine sozial- und kulturwissenschaftliche Einführung in die Ernährungsforschung* [Sociology of Food: a Social and Cultural Introduction to Nutrition Research], Weinheim, 1999.

Barlösius, E. 'Der ewige Streit über die "richtige" Ernährung. Auswirkungen auf Produktion, Konsum und Politik' [The Never-Ending Dispute over the "right" Diet: Impact on Production, Consumption and Politics], in Bund für Lebensmittelrecht und Lebensmittelkunde (eds) *Sachen Lebensmittel. Jahrestagung '99. Ansprachen und Vorträge*, Bonn, 1999/2000, 51–8.

Barlösius, E. 'Ernährungsziele – Ein Kommentar aus Sicht der soziologischen Essforschung' [Nutrition Goals: a Sociological Commentary], in Oltersdorf and Gedrich (eds) 2001, 113–25.

Benterbusch, R. *Inhaltsanalyse zum Thema Ernährung in deutschen Zeitungen (1994/95)* [Content Analysis of Nutrition in German Newspapers], Karlsruhe, 1997.

Borscheid, P. and Wischermann, C. (eds), *Bilderwelt des Alltags. Werbung in der Konsumgesellschaft des 19. und 20. Jahrhunderts* [Images of the Everyday World: Advertising in the Consumer Society of the Nineteenth and Twentieth Centuries], Stuttgart, 1995.

Burnett, J. and Oddy, D.J. (eds), ICREFH II, 1994.

Dannenberg, A. 'Ernährungs- und Gesundheitskampagnen' [Food and Health Campaigns], in Oltersdorf and Gedrich (eds) 2001, 77–8.

Deutsche Gesellschaft für Ernährung im Auftrag des Bundesministeriums für Jugend, Familie und Gesundheit und des Bundesministeriums für Ernährung, Landwirtschaft und Forsten (eds) [German Society for Nutrition on Behalf of the Federal Ministry for Youth, Family and Health and the Federal Ministry of Food, Agriculture and Forestry], *Ernährungsbericht 1976* [Nutrition Report], Frankfurt am Main, 1976.

Deutsche Gesellschaft für Ernährung im Auftrag des Bundesministeriums für Verbraucherschutz, Ernährung und Landwirtschaft (eds), *Ernährungsbericht 2004* [Nutrition Report], Bonn, 2004.

Fröhlich, S. *Das Moppel-Ich. Der Kampf mit den Pfunden* [Fighting the Flab], Frankfurt am Main, 17th ed., 2004.

Habermas, T., *Zur Geschichte der Magersucht. Eine medizinpsychologische Rekonstruktion* [The History of Anorexia: a Medical-Psychological Reconstruction], Frankfurt am Main, 1994.

Herden, B. 'Einfach essen' [Just Eat], *Die Zeit*, 46, 9 November 2006.

Hersey, G.L. *Verführung nach Maß. Ideal und Tyrannei des perfekten Körpers* [Temptation to Measure. Ideal and Tyranny of the Perfect Body], Berlin, 1998.

Lorenz, M., *Leibhaftige Vergangenheit. Einführung in die Körpergeschichte* [Introduction to the History of the Body], Tübingen, 2000.

Merta, S., *Wege und Irrwege zum modernen Schlankheitskult. Zur Diätkost und Körperkultur als Suche nach neuen Lebensstilformen 1880–1930* [Paths and Blind Alleys to the Modern Cult of Slenderness: for Dietary and Physical Culture as a Search for New Lifestyles, 1880–1930], Stuttgart, 2003.

Oddy, D.J. and Petranova, L. 'The Diffusion of Food Culture' in ICREFH VIII, 18–28.

Oltersdorf, U. and Gedrich, K. (eds), *Ernährungsziele unserer Gesellschaft: Die Beiträge der Ernährungswissenschaft* [Nutrition Goals of our Society: the Contributions of Food Science], Karlsruhe, 2001.

Rath, C-D. *Reste der Tafelrunde. Das Abenteuer der Eßkultur* [Remains of the Round Table: the Adventure of Food Culture], Reinbek bei Hamburg, 1984.

Sarasin, P. and Tanner, J. (eds), *Physiologie und industrielle Gesellschaft. Studien zur Verwissenschaftlichung des Körpers im 19. und 20. Jahrhundert* [Physiology and Industrial Society: Studies on the Scientization of the Body in the Nineteenth and Twentieth Centuries], Frankfurt am Main, 1998.

Selter, B. '"Der satte" Verbraucher: Idole des Ernährungsverhaltens zwischen Hunger und Überfluß 1890−1970' [Rich Consumers: Models of Nutritional Behaviour Between Hunger and Abundance 1890−1970], in Borscheid and Wischermann (eds), 190−221.

Schmidt, J., 'How to Feed Three Million Inhabitants: Berlin in the First Years after the Second World War', in ICREFH IX, 63−73.

Schwartz, H. *Never Satisfied: a Cultural History of Diets, Fantasies and Fat*, New York, 1986.

Spiekermann, U. 'Vollkorn für die Führer. Zur Geschichte der Vollkornbrotpolitik im "Dritten Reich"' [Wholegrain for the Führer: the History of Wholemeal Politics in the "Third Reich"], *Zeitschrift für Geschichte des 20. und 21. Jahrhunderts* 16, 2001, 91−128.

Spiekermann, U. 'Historischer Wandel der Ernährungsziele in Deutschland − Ein Überblick' [Historical Change of Diet Goals in Germany: an Overview], in Oltersdorf and Gedrich (eds), 2001, 97−112.

Tanner, J. 'Der Mensch ist was er ißt. Ernährungsmythen und Wandel der Eßkultur' [Man is What he Eats: Nutrition Myths and Changing Food Culture], *Historische Anthropologie* 4, 1996, 399−419.

Thoms, U. 'Körperstereotype. Veränderungen in der Bewertung von Schlankheit und Fettleibigkeit in den letzten 200 Jahren' [Body Stereotypes: Changes in the Valuation of Thinness and Obesity in the Last 200 Years], in Wischermann, C. and Haas, S. (eds), *Körper mit Geschichte. Der menschliche Körper als Ort der Selbst- und Weltdeutung*, Stuttgart, 2000, 281−307.

Teuteberg, H-J. (ed.), *Revolution am Esstisch. Neue Studien zur Nahrungskultur im 19./20. Jahrhundert* [Revolution at the Dining Table: New Studies of Food Culture in the Nineteenth and Twentieth Centuries], Stuttgart, 2004.

Werner, P., *Vitamine als Mythos. Dokumente zur Vitaminforschung* [Vitamins as a Myth: Documents on Vitamin Research], Berlin, 1998.

Wickum-Glinski, S., *Die Darstellung des menschlichen Körpers im Kinder- und Jugendbuch von der Aufklärung bis zur Gegenwart* [The Portrayal of the Human Body in Children and Young People from the Enlightenment to the Present], Herzogenrath, 1998.

Zweininger-Bargielowska, I., 'Building a British Superman: Physical Culture in Interwar Britain', *Journal of Contemporary History* 41, 2006, 595−610.

Chapter 12

Food Consumption and Risk of Obesity: The Medical Discourse in France 1850–1930

Julia Csergo

While excess weight and its associated health risks have been at the forefront of the concerns of medicine since ancient times, it was not before the mid-nineteenth century that the issue of obesity became particularly important for medical research, which was undergoing something of an upsurge at the time. There was an increase in the number of works on obesity, its clinical description, pathogenesis and therapy. Alongside the medical discourse was a gastronomic discourse which focused more on the general relationship between food and health. Beginning with Brillat-Savarin, who devoted one of the meditations in his *Physiologie du Goût* to the issue of obesity, the movement culminated in the decades between 1910 and 1930 with cookery books by authors such as Henri Babinski (pseud. Ali-Bab) and Paul Reboux, and it even worked its way into literary discourse.[1] The short story by Henri Béraud, *le Martyre de l'Obèse*, the Prix Goncourt winner in 1922, led in 1932 to a response by Dr Hemmerdinger in the form of *La Fin du Martyre de l'Obèse*.[2] Meanwhile, obesity was coming to be construed as a pathology of the century of abundance and modernity.

Obesity, a Notion under Construction

Defined as 'a pathological condition caused by the general hypertrophy of adipose tissue,' obesity was distinguished from excess weight, 'the condition of the body of people who are fat' and 'polysarcia,' defined as extreme obesity.[3]

While today norms are calculated in relation to the Body Mass Index (BMI), for many years this measurement was not used and the assessment of obesity was based on a visual analysis which was entirely relative – an increase in the volume of the body, and change in the individual's appearance. A distinction was also made between being overweight and 'excessive overweight which degenerates into obesity.'[4]

---

1 Brillat-Savarin 1826, Ali-Bab 1907, Reboux 1931.
2 Béraud 1922, Hemmerdinger 1932.
3 Deschambre 1880.
4 Dubois 1912.

However, despite part of the medical establishment having reservations in this regard, there were attempts throughout the nineteenth century to introduce more scientific assessments on the basis of a relationship between weight and size in order to establish a reference norm. Before Quetelet, the dominant viewpoint was that a normal man should weigh, at a given age, 'as many kilograms as his size in centimetres above one metre.'[5] According to this principle, on average, a man of 1.70 m [5 foot 7 inches] should weigh 70 kg [11 stone] (or, if we translate this relationship into BMI, the norm of the years 1850–1930 gives a BMI of 24.2, which is close to the upper limit or normal body weight today). There were, however, variations that took into account sex, complexion and musculature. These led to mathematical formulae that incorporated a leeway of a few kilos.[6] Between the years 1870 and 1935, standards for men, in BMI terms, varied from 22 to 27, i.e. a range from normal (18.5 to 24.9) to overweight (25 to 29.9).[7] It was only beyond that level that obesity began so that until the 1930s the present range of being overweight was considered as within the bounds of normality. The threshold which defines obesity is therefore a cultural variable.

The Aetiology of Weight Growth

Fat is part of the composition of the human body: it represents about 10 per cent to 20 per cent of body mass.[8] A normal man weighing 70 kg has 3 kg or 4 kg of fat reserves, which are held in the folds of adipose tissue. Obesity is an abnormal increase of this adipose tissue and an accumulation of fat in the body as a whole such that the functioning of the organs is affected. Hence the only definition given in the nineteenth century was that 'obesity begins when there is functional impairment.'

The first studies in France specifically focusing on obesity were those by Dr J.F. Dancel who looked at the development of fat in the human species (1852), and experimented, using the principles of organic chemistry, in order to reduce excess weight. In 1863, he contributed the first major work on obesity, or 'excessive portliness', which he defined as an abnormal condition associated with numerous pathologies.[9] Later, works by Drs Dubourg and Worthington determined the complex origins of this condition and its aetiology, due to functional disorders affecting the regulation of fat metabolism. For example, they underlined that the causes were scrofulous, cardiac, nervous – the influence of emotions – and endocrine – thyroid, pituitary, genital and adrenal deficiency. They also indicated that liver sclerosis and syphilis, for example, and pancreatic insufficiency, as

---

5   Dr Broca's Index. Quetelet 1870.
6   Ramond 1907.
7   Raffray 1912, Heckel 1911, Feuillade 1935.
8   Dechambre 1880.
9   Dancel 1863.

well as fatigue, defined as an aspect of auto-intoxication, could be implicated.[10] Hereditary predispositions, whether direct or collateral, were highlighted by studies showing that, in about 50 per cent of cases, obese parents had obese children and the doctors focused on obese families with 'successions of gout sufferers, diabetics and neuropaths'.[11]

It was not until the work of Professor Bouchard, from 1874 onwards, that a first attempt was made to formulate a scientific definition of obesity: in his view, the reason that fat accumulated in certain subjects was that they suffered from decelerated combustion and/or from dissimulation (when food produces fat to the detriment of protein). It was not therefore because an individual was too fat that he was said to be 'obese,' but because he conserved fat abnormally or because his body produced excessive fat.[12]

Is the Cause of Obesity to be Found in Food?

Once this aetiology was acknowledged, the question turned to whether obesity was caused by food intake; in other words whether a direct production or accumulation of fat in the body could occur as a result of the ingestion of food. In 1826, Brillat-Savarin had already described the diet of obese people as being too rich in flour and starch components, since their appetite was for bread, rice, pasta, potatoes, starch (beans, etc.) and various flours, fat, sugar and therefore pastries.

Food hygiene specialists turned their attention to the newly developing food science. Beginning with the work of Lavoisier, Liebig and secondarily Dumas, mid-nineteenth century chemistry was able to determine, by analysing the chemical composition of foodstuffs and of the modifications they underwent in the body, a new categorization of nutrients: distinctions were drawn between 'albuminoid' or protein foods (meat, egg white, milk casein, cereal gluten and bread) whose role is to repair tissue; fats (lard, leaf fat, egg yolk, butter, milk or oils) which produce heat (important in colder climates); and carbohydrates (vegetable starch, potatoes, sugars, fruit), energy producing combustible food.

It was the experiment conducted by Flourens which paved the way for research on the role of food in the increase of fat. In 1843, he observed the stoutness of the bears in the Botanical Gardens, fed exclusively on bread. At the same time, Boussingault, Dumas and Payen worked on the role played by fat in weight increase, demonstrating the dual process which affects fat when it is absorbed by the body: first, it burns to provide heat and energy; second, it fixes itself in the tissues to provide a reserve for respiration. In 1845, Boussingault went a step further by holding that fatty foodstuffs were not solely responsible for the formation of fat within the body: carbohydrates could also turn into fat (proof was not provided

---

10   Dubourg 1864, Worthington 1877.
11   Chambers 1850, Worthington 1877, Raffray 1912.
12   Feuillade 1935.

before Soxhlet in 1881) as can the 'albuminoid' substances or proteins. This latter hypothesis was only generally acknowledged at the turn of the century.[13] Thus, eating too much meat or game could be a factor in obesity. Other obesity sufferers ate very little meat and had a diet consisting almost exclusively of pulses, potatoes, starches and bread, especially fresh bread. Fats, like lard, butter, oil and cold pork meats and carbohydrates such as sugar and starch, the favourite foods of the average obesity sufferer, also formed fat; it is the association of these various foodstuffs which encouraged fast weight growth.[14] It was long thought that the absorption of liquids played an essential role in increasing the volume of the body and the swelling of tissues. Dancel, in this respect, called for a severe restriction of water intake for the obese until Debove, in 1885, demonstrated that the volume of water absorbed by an individual does not affect body weight. By contrast, alcohol, wine, beer and spirits were recognized as favouring excess fat growth.

Given the complexity of the situation, how was it possible to determine the quantity of food beyond which the body begins to develop fat? While Dr Javal advocated slimming diets involving a certain volume of food in order to assuage hunger, the attention of the medical community began to focus on the issue of the energy balance: for example, there was no need for a slimming diet to take any account of the fat, protein or carbohydrate content of the foodstuffs, but simply to focus on the energy value of the diet in question.[15] It was on this basis that the link between obesity and food consumption was built, one which still tends to prevail in the modern era.

Obesity and the Excess Consumption of Food: The Case of the 'Big Eater'

Not all obesity is due to food consumption. Bouchard's statistics (1876) were still very much a reference point in the 1930s, showing that out of 100 obesity sufferers (all categories taken together) 50 had a normal diet, 40 were big eaters and ten had an insufficient intake. Ramond, in 1907, confirmed this finding when he demonstrated that out of the 65 obese patients observed, 30 had a normal appetite, 28 an excessive appetite, seven a limited appetite. Brillat-Savarin himself saw excessive eating and drinking as only one of the causes, among others, of obesity.

In order to establish a link between obesity and food consumption it is necessary to consider the big eater. At first glance, he can be recognized by a form of obesity limited to the belly, because of eating too much. But how is this excess food intake to be measured? Does it mean eating beyond one's hunger? But does the sensation of hunger itself express organic needs? One of the privileges of the human species, according to Brillat-Savarin, is 'eating without hunger and drinking without thirst.'

---

13   Leven 1900.
14   Ramond 1907.
15   Javal 1900.

It is misleading to focus on the hunger of someone who eats with voracity and insatiability as if intoxicated, which leads to an ever-increasing food intake and excites the motor and secretory functions of digestion.

Overeating, a so-called 'exogenous' or 'occasional' cause of obesity, has been defined as the provision of a quantity of food in excess of needs, something which facilitates fat increase among subjects who are sedentary and predisposed.[16] However, medical opinion was not unanimous: while the Dechambre dictionary states that polysarcous, or fleshy, individuals are generally big eaters – 'they spend, either by habit, or out of real necessity, a considerable portion of their time eating or drinking,' for many others, overeating can under no circumstances lead to polysarcia (fleshiness) in a healthy subject. In 1904, Dr Leven even said that overeating never led to considerable weight gain on its own, since a person's weight remained more or less as invariable as their temperature. And in 1935, Dr Feuillade went as far as to assert that 'only those subjects who become fat in the absence of any overeating should be considered as obese.'[17]

However, the incrimination of overeating, something which still predominates today, took much of its impetus from the establishment of the new discipline of food science, which made it possible to assess the energy value of foodstuffs on the basis of the quantity of energy which its combustion releases within the body and therefore to express in terms of kilocalories or, recently, as kilojoules, the amount of energy the body is obliged to expend under various life conditions.

The assessment of the average energy expenditure of the body, estimated during 24-hour periods, as 300 g of carbon (air, excrement, urine), 20 g of nitrogen (urea, uric acid) and 30 g of minerals (2 to 3 litres of water) led to the notion of food requirements being developed – that is, the amount of materials necessary to renew worn tissues, provide energy necessary for work, and maintaining the body temperature at 37° Celsius. These amounts took into account the work performed by the subject, or effort expended, stature and weight, various physical states (growth, pregnancy, old age) and the external environment (climate, season). A distinction was therefore made between a working diet which, for the purpose of physical work required carbohydrate-rich food in which sugar must predominate, and a maintenance diet, necessary for repairing tissue and producing heat. These two intakes combined were expressed in kilocalories (kcals), for various different cases. For example, a man of 60 kg was said to need a daily intake of 2,250 kcals if he led a sedentary life (office worker), 2,400 kcals if he was engaged in moderately energetic work (carpenter, bricklayer) and 2,880 kcals if he was undertaking strenuous activity (docker, road builder). As for the amount of proteins necessary, it was fixed at 1 gram per kilogram of bodyweight (or 70 g of meat per day for a man of 70 kg). If the intake of energy exceeded that expended, the excess formed fat reserves.

---

16   Dechambre 1880.
17   Feuillade 1935: 10.

Being overweight due to overeating differs from the pathology of obesity. In this respect, Dr Feuillade differentiated between confusing:

> overweight subjects, who may be normal, with obese people, who are to be regarded as patients...a well fed individual has a level of fat which is by no means pathological, if his body is working regularly and enables him to lead a normal existence. An increase in food intake and regular exercise will enable him to use up the fat reserves and return to a weight which corresponds to his size.[18]

By demonstrating the way that overeating functioned, the medical profession drew attention to the disorders generated by an increase of fat – the fattening of organs such as the heart or liver, renal impairment, sugar, iron, urea and nitrogen assimilation pathologies, disorders that in turn were liable to engender the pathogenesis of obesity. A simple obesity, one caused by overeating, could therefore be turned into a more complicated form of obesity.[19]

The Historical Sociology of the 'Big Eater'

Eating too much is nowadays behaviour associated first and foremost with the poorer socioeconomic classes.[20] This does not seem to have been the case in the nineteenth and early twentieth centuries. During this period, popular imagery associated the 'fat man' with excessive food consumption made possible by wealth and was therefore in itself a symbol of this wealth.[21] In literary and pictorial representations, the 'fat man' was an embodiment of somebody who was rich, a landowner with power, all of those attributes commonly associated with gastronomic circles. Before 1914, as Heckel asserted, 'overeating and inactivity are the norm among bankers, politicians, butchers, the confectioners and motorists.'[22]

Excluding cases of morbid obesity, it would appear that the 'obesity of the belly' had very little connection, during this period, with the most disadvantaged sections of the population. Beyond the representations analysed by Jean-Paul Aron, there are only very limited data on the eating practices of ordinary people: the autobiographies of workers reveal a frugal and scarce diet; as for the surveys undertaken by socially-conscious doctors, it is clear that they tended to highlight the insufficiency – both quantitative and qualitative – of food rather than excess food intake.[23] Taking examples from the papers presented at the First Scientific

---

18   Feuillade 1935: 6.
19   Labbé 1905.
20   Poulain 2002.
21   Aron 1973, Brillat-Savarin 1826.
22   Heckel 1931.
23   Lhuissier 2007.

Congress of Food Hygiene held in 1904, they show that, until the early twentieth century, nutritional issues focused mainly on the inadequacy and irrationality of workers' diets, in particular with regard to the pathogenesis and prophylaxis of tuberculosis, which at the time was responsible for more than 10,000 deaths a year in France.[24] Tuberculosis was the principal infectious disease of the nineteenth century, and was the main focus for the hygienists and the public authorities.[25] In 1852, Professor Bouchardat had explained the development of tuberculosis among the poor by the lack of food and the lack of heat produced by the body, hence the research conducted on the nutritive power of meats and the creation of meat concentrates and extracts.[26] At the 1904 Congress, Professor Armand Gautier claimed that 'The most authoritative works have recently established that this terrible ill is contracted not through the dust introduced into the lungs but through the food which enters the digestive tract.'[27]

Several surveys demonstrated the insufficiency of workers' diets. The study conducted by Dr A. Imbert monitored the diet of an individual subject through the various different stages of his professional life.[28] The subject was a postman who began his working day at five o'clock in the morning and continued until ten o'clock in the evening with five hours of combined rest periods during the day. His breakfast consisted of a black or white coffee, a bar of chocolate or a dry sausage and a small hunk of bread, while at noon and in the evening, he had soup, a little bread, 70 g of meat, some vegetables, fruit or cheese. Transferred to a postman's position in the Vaucluse, he woke up at four o'clock in the morning, travelled two kilometres to reach his work, and then between 32 and 35km to deliver his post. The 14 main meals of the week were fairly evenly distributed as shown in Table 12.1. The lunchtime meals comprised mostly fats and starchy foods, which produced heat and energy, and the evening meals comprised protein and carbohydrates, for tissue repair and energy production. This diet was said to be balanced because it corresponded to the major energy expenditure required by the subject's professional occupation.

Another survey was presented by Dr Labbé, an activist for physiological and rational diets for workers that were nutritive, safe and economic. He looked at the advice given at the Laennec hospital in Paris to hospital patients and those entitled to free consultations who suffered from tuberculosis, moderate respiratory conditions and syphilis.[29] In this survey, the doctor did not provide details of the diets of the hundred or so patients surveyed, but he supplied the conclusions he drew from his observations of them. The results in Table 12.2 demonstrated that their diets were insufficient and unbalanced.

---

24 Gauthier 1904.
25 Bardet 1998.
26 Csergo 2004.
27 Gauthier 1904.
28 Imbert 1904.
29 Labbé 1905.

Table 12.1   Consumption of Foods by a French Postman, 1904

Foods used	Distribution over seven lunches (times/week)	Distribution over seven dinners (times/week)
Fat	3	1
Oil	3	2
Cheese	2	-
Potatoes	5	2
Dried beans, lentils	4	2
Pasta	-	3
Salad, fresh vegetables	1	2
Fruit	3	6
Meat	-	6

Source: Imbert, 1905.

Table 12.2   Diets of Patients Attending a Paris Hospital, 1905

	Too much	Not enough	Ration consumed (kcals)	Ration necessary (kcals)
Worker performing strenuous tasks	Meat Alcohols	Vegetables Pasta Starches Sugars	4,600	3,600
Worker performing moderate tasks	Meat Liquor, wine	Vegetables, soups Pasta Starches Sugar, sweet foods	2,400	2,600
Sedentary worker	Everything, especially meat Aperitifs, liqueurs	Fresh vegetables Sweet dishes Water	3,200	2,100
Parisian workers and employees	Low energy food, condiments (lettuce, radishes, raw vegetables, poor quality fruit, pickles, salad dressing)	Bread Meat Starches Pasta Soups Sweet dishes	1,400	2,100

Source: Labbé, 1905.

Dr Labbé went on to establish an assessment of the dietary preferences of Parisian workers. In Table 12.3 these are ranked in descending order. For men, we can see the importance of wine – which doctors recommended should be consumed in lesser amounts – and also of liqueurs which it was recommended should be consumed in very small amounts. Bread was the staple food par excellence, both cheap and nourishing, and coffee the ideal and highly recommended 'nervine' foodstuff. Pulses, presented as reservoirs of energy, were little eaten, despite being healthy and inexpensive if substituted for meat; soups were inexpensive and comforting; milk was generally consumed in accordance with dietetic recommendations.

Table 12.3   Percentage of Interviewees Consuming Certain Products Daily, 1905

Men		Women	
	(%)		**(%)**
Red wine	86.6	Fresh vegetables	91.5
Bread	80.0	Bread	69.5
Coffee	75.5	Pasta	61.7
Soup	75.5	Coffee	61.6
Fresh vegetables	73.3	Milk	61.5
Aperitifs	71.4	Soup	61.1
Meat	68.8	Beer	55.9
Milk	66.6	Water	55.9
Pasta	61.1	Meat	52.5
Liqueurs	60.0	Red wine	50.8
Beer	48.8	Pulses	37.2
Pulses	40.0	Pastries	30.5
Water	31.0	Liqueurs	23.7
White wine	28.5	Sugar	20.3
Sugar	24.4	White wine	10.1
Pastries	17.7	Aperitifs	7.1

Source: Labbé 1905.

However, certain observations with regard to these diets may seem paradoxical to modern eyes. They demonstrate how dietary habits have changed over a century, which must be linked to changes in work patterns and the gradual increase in sedentary labour. In his conclusions, Dr Labbé deplored the excessive consumption of fresh vegetables and fruit, seen as a fault in the diet of women, because they were expensive and were not comforting; the under-use of pasta which the doctor would have preferred to have seen more widely used because of its excellent nutritional value and the fact that it was inexpensive; the low consumption of sugar – sweets, confectionery and chocolate – associated with the folk idea that it was not a food but a condiment, and which the doctor recommended should form part of the daily diet because of its high energy, nutritional and economic qualities; the 'regrettable lack of appetite' for pastries and sweet dishes, the energy value of which was four times higher than that of sirloin steak.[30] Regarding meat, Dr Labbé indicated that it was an ideal food but stated his belief that the proportion of energy it provided did not always justify the high cost involved.

Another survey, conducted by J. Tribot (1904–5), looked at the energy value of food served in Parisian *Crémeries*, soup kitchens and fixed-price restaurants.[31] The study reminds us that fast-food commercial catering has always been a constant feature of workers' diets, and underlines the insufficiencies associated with this kind of food. The standard meal on offer in the *Crémeries*, where the office clerks and workers in the sewing and fashion trades ate, was generally made up of two fried eggs, half a litre of milk or a cup of chocolate or white coffee, a small piece of bread and between two and five lumps of sugar. The survey findings were clear: if the subject ate at the *Crémerie* noon and evening, the nutritional value of the meals was well below what was required, despite a slight surplus of carbohydrates: 87g of protein instead of the 100 to 125g required; 54g of fat instead of the 60 to 80 g needed; 116 g of carbohydrates instead of the 400 to 450g needed. The meals on offer in the fixed-price restaurants (1.15 to 1.25 francs per meal) or the soup kitchens (frequented by workers) seem to have been more advantageous in terms of nutrition, being made up of as much bread as was wanted, a meat and vegetable dish, cheese or pastry, white or red wine. For this outside commercial catering, where workers and clerks ate on a daily basis and where the food choices took their inspiration from 'frivolous and irresponsible' appetites rather than the reasoned assessment of the nutritional value of the dish of the day, Tribot suggested, in order to calculate the nutritional value of the meals, that 'each person calculate his own food'. He proposed to make a set of scales available for regular customers, with a printed table setting out the weight from 40 to 90 g and the corresponding calorie-values necessary for physiologically appropriate food intakes. Dietary indicators of nutritional values were to be passed around like menus.

The 'Fat Invasion,' and the Mismanagement of Abundance

From 1910 onwards, while popular representation continued to associate 'fat' with the rich and powerful, the medical discourse on obesity underwent a transformation.

---

30   Bruegel 2001, Csergo 2008.
31   *Revue Société Scientientifique d'Hygiène Alimentaire* 1904: 56–69.

This change was heralded first of all in Germany in the late nineteenth century, before becoming noticeable in France, where the German health policy began to attract attention.[32] The pathogenesis of obesity was well known by this time and major work continued to establish the link between obesity and endocrinology. Treatments varied and, while taking into account the pathological condition leading to weight increase, began to move towards working in three different domains: the psychic (going about one's occupations, entertainment, travelling to take one's mind off food), the physical (exercises and skin stimulation, baths, massages) and the diet (appropriate food intake). But at a time when thinness was 'more fashionable than ever', digestive obesity, i.e. the big eater, continued to form an integral part of the discourse.[33] This was because, in conjunction with the scientific and technical progress that had been made since the beginning of the industrial age, major transformations had occurred since the 1880s in the field of food production, thus signalling an end to the era of 'shortage' and opening an age of abundance for all. The new conditions of supply, industrial mass production – in particular that of 'inferior similars' or substitute foods for the masses – and volume retailing led to decreasing food prices and mass consumption. This process was dubbed 'the levelling out of enjoyment' by Viscount d'Avenel in 1913.[34] Between 1850 and 1914, figures for the consumption of meat, sugar and fats – of inferior quality such as margarine – show a clear progression.[35]

This movement, the pathogenic abuses of which were amply denounced, went from strength to strength after the Great War and heralded a period in which overeating became much decried in France.[36] A doctor such as Dr Carton, a defender of 'synthetic, scientific and philosophical naturism,' saw overeating both as a sign and a factor of degeneration and of individual and social neurosis being imminent.[37] He held generalized overeating accountable for the increased frequency of obesity, madness (14,900 in 1865, 71,547 in 1910), suicide (1,739 in 1830, 9,619 in 1909), arthritis, tuberculosis, the sufferers of which could be thin or fat depending on the form of their disease – and cancer (30,000 deaths per year). Harking back to a time when peasants ate black bread, vegetables and fresh water, a period of 'lost robustness', he singled out the intensive culture of modern societies based on the consumption of meat, sugar and alcohol – these 'three murderous foods' – which he held responsible for the decrepitude of bodies. The development of the mass market of foods which had been devitalized or created by industry, such as canned foods, sucrose or industrial fats, was also incriminated.[38]

---

32 Thoms 2007.
33 Heckel 1931.
34 Avenel 1913. 'Inferior similars' are industrial products made from cheap raw materials enhanced by colouring agents, texturizing agents, or aromas. See Csergo 2004.
35 Csergo 2004.
36 Pascault 1906.
37 Carton 1912, Drouard 1998.
38 Allendy 1926.

Dr Heckel, a staunch advocate of vegetarian diets and a specialist in obesity and its treatment, also noted that between 1914 and 1930, food intake had increased by 30 to 200 per cent depending on social class, with the steepest progression being observed among rural and working-class populations.[39] The numbers he propounded, which it must be admitted are difficult to verify, show a considerable increase in food intake which, for the grand bourgeoisie and 'those made rich by the war' had increased from 3,000–3,500 kcals [12,500–14,600 kj] per day before 1914 to 4,500–6,000 [18,800–25,000 kj] after 1914, and for workers and peasants from 2,000–2,500 kcals [8,400–10,500 kj] per day before 1914 to 4,000–4,500 [16,700–18,800 kj] after 1914. He also denounced the overeating observed among three-quarters of the population in the 'bankrupt' nations, and the excessive consumption of proteins, in particular among peasants and working populations who now ate meat on a daily basis, to the detriment of fruit and vegetables. 'Workers and farmers now suffer from gout, obesity, diabetes, neurasthenia, angina pectoris and cerebral haemorrhages, pathologies formerly reserved for the rich and for aristocrats,' said Dr Heckel, deploring the ruinous effect on health of overeating, dietary debauchery and 'general gluttony': the 'folly of eating'. This digestive frenzy, which in the nineteenth century only affected a small and wealthy elite, now extended to the whole of society, as the huge increase in the number of restaurants and culinary books and magazines attested: 'It would seem that stuffing oneself is the major concern of the great capitals,' bemoaned Dr Heckel. Better still, he went on to assert:

> among the industrialists and traders who arose during the war, thousands of people of common extraction, with neither education nor culture, have gained access to considerable fortunes ... the sign of success and wealth lies in the possibility of eating without restraint and of having an abundance of the most highly sought after foods ... the desire to make up for former deprivations, to feast.[40]

Overeating and 'obesity of the belly' now appeared as a 'defect of civilization,' a mismanagement of wealth and abundance. Carton presented it as follows:

> Every time a civilization comes into possession of wealth above its strict requirements, this well-being, instead of being used in accordance to natural laws, is dissipated in order to infringe them, i.e. to gorge on meat, to stuff oneself with sweetmeats, to get drunk on fermented beverages and to deprive oneself as far as possible of every opportunity to undertake muscular effort ... in the final analysis, wealth, with its consequences of anti-physiological diet and sedentariness, is the major source of disease and degeneration.[41]

---

39   Heckel 1931.
40   Heckel 1931: 56.
41   Carton 1912: 53.

Socially Differentiated Medical Discourses

In light of the risks of overeating, the early twentieth century saw the establishment of a number of socially differentiated discourses recommending improved management of abundance and the need to combat 'obesity of the belly'. In the tradition of Ali-Bab, the pseudonym of the distinguished chef Henri Babinski, whose hugely successful cookbook 'Practical Gastronomy' was republished continuously between 1907 and the 1950s and was enriched by an extra chapter entitled 'Treatment of the obesity of the greedy', using a medical discourse with 'gastronomic' undertones which aimed to combat the obesity of the elites. Here again, Heckel, who, strongly against what he called the 'scams of obesity' − the industry of massages and electrical gymnastics − offered slimming diets based on pleasant, varied and carefully prepared cuisine, 'pleasure food':

> If you prescribe eggs, there is no reason why they should not be scrambled with truffles or mushrooms. There is no need for a sole to be dried or boiled, it can perfectly well be embellished. Green vegetables in chicken stock are as delicious as with butter.[42]

The patients Heckel was addressing were also socially identified by the energy infringements of their slimming diet: 'the occasional and restricted enjoyment of a glass of champagne or a small quantity of very pleasant or highly stimulating wine (port), or an extremely flavoursome dish (foie gras, goose preserve, cassoulet, bouillabaisse).'[43] Dr Paul Reboux, wrote a similar treatise, on 'new diets', in which he presented 300 recipes which complied with medical rules but which were combined in accordance with gastronomic precepts, such as taste and sensory pleasure.[44]

For ordinary folk, the question was a different one: here, obesity appeared to be due to pathogeneses other than overeating, which were only just beginning to emerge among the less affluent. The food risk that was denounced concerned first and foremost the poor nutritional quality of industrial food available, perhaps adulterated or ersatz, the excessive consumption of alcohol and meat and the over consumption of vegetables and fruit by women. The medical discourse never took the gastronomic aspect on board, and so the pleasure of food was never a concern. The medical discourse focused mainly on nutrition − an awareness of the nutritional and energy values of food − and the necessity of food hygiene and rational diet education, in particular through the implementation of food education both by food manufacturing companies and in the home-economics classes of the schools of the Republic.

---

42 Heckel 1911.
43 Heckel 1931.
44 Reboux 1931.

References

Allendy, R. *Précis de Thérapeutique Alimentaire*, Paris, 1926.

Ali-Bab (Henri Babinski) *Gastronomie Pratique*, Paris, 1907.

Apfeldorfer, G. *Maigrir, c'est Fou*, Paris, 2000.

Aron, J-P. *Le Mangeur du XIXe Siècle*, Paris, 1973.

Avenel, G. d' *Le Nivellement des Jouissances* [The Levelling of Enjoyment], Paris, 1913.

Bardet, J.P. (ed.), *Peurs et Terreurs Face à la Contagion* [Fear and Terror in the Face of Contagion], Paris, 1988.

Béraud, H. (ed.), *Le Martyre de l'Obèse* [The Martyrdom of the Obese], Paris, 1925.

Bouchard, C. *Maladies par Ralentissement de la Nutrition* [Deseases from Slowing Down of Nutrition), Paris, 1882.

Brillat-Savarin, J.A. *Physiologie du Goût ou Méditations de Gastronomie Transcendante* [The Physiology of Taste: or, Meditations on Transcendental Gastronomy], Paris, 1826.

Bruegel, M. 'A Bourgeois Good? Sugar, Norms of Consumption and the Labouring Classes in Nineteenth Century France', in Scholliers, P. (ed.), *Food, Drink and Identity: Cooking, Eating and Drinking in Europe since the Middle Ages*, Oxford, 2001.

Carton, P. *Les Trois Aliments Meurtriers* [Three Deadly Foods], Paris, 1912.

Chambers, T.K. *Corpulence or Excess of Fat in the Human Body*, London, 1850.

Csergo, J. 'Entre faim légitime et frénésie de la table au XIXe siècle: la constitution de la science alimentaire au siècle de la gastronomie' [Between genuine and frenzied hunger at the nineteenth century table], 2004. Available at: http://www.lemangeur-ocha.com/fileadmin/Pdf_agenda_et_actus/faim_legitime_frenesie_table_int.pdf [accessed 22 December 2008].

Csergo, J. 'Les Mutations de la Modernité Alimentaire' [Evolution of the Modern Diet], in *Des Aliments et des Hommes: Entre Science et Idéologie, Définir ses Propres Repères, Actes du Colloque de l'Institut Français de Nutrition, 8 et 9 Décembre 2004*, Paris, 2005.

Csergo, J. 'Le Sucre: de l'Idéalisation à l'Ostracisme' [Sugar: from Idealization to Ostracism], *Cahiers de Nutrition et de Diététique*, series 2, 43, 2008.

Dancel, J.F. *Traité Théorique et Pratique de l'Obésité* [A Theoretical and Practical Treatise on Obesity], Paris, 1863.

Deschambre, A. *Dictionnaire Encyclopédique des Sciences Médicales* [Encyclopedic Dictionary of the Medical Sciences], Paris, 1865–80.

Drouard, A. 'Le Régime Alimentaire du Dr Carton' [The Dietary Regime of Dr Carton], *Cahiers de Nutrition et de Diététique*, 33, 2, 1998.

Dubois, R. *Comment Maigrir. Moyens Efficaces, Conseils Pratiques et Régimes pour Vaincre l'Excès d'Embonpoint* [How to Lose Weight: Effective Tips and Regimes for Overcoming Excessive Overweight], Paris, 1912.

Dubourg, L. *Recherches sur les Causes de la Polysarcie* [Research on the Causes of Obesity], unpublished medical thesis, Paris, 1864.

Feuillade, H. *Le Livre de l'Obèse* [The Book of Obesity], Vichy, 1935.

Gauthier, A. Introduction, *Revue de la Société Scientifique d'Hygiène Alimentaire et de l'Alimentation Rationnelle de l'Homme* 1, 1904, xv.

Revue scientifique d'hygiène alimentaire et de l'alimentation rationnelle de l'homme, Paris, 1904–1909.

Heckel, F. *Grandes et Petites Obésités* [Major and Minor Obesity], Paris, 1911.

Heckel, F. *Maigrir. Pourquoi? Comment? Conceptions et Méthodes Nouvelles du Docteur Francis Heckel* [Lose weight. Why? How? New Concepts and Methods of Dr Francis Heckel], Corbeil, 1931.

Hemmerdinger, A. *La Fin du Martyre de l'Obèse* [The End of the Martyrdom of the Obese], Paris, 1932.

Imbert, A. 'Une Observation Économique de la Vie Ouvrière' [An Economic Comment on the Worker's Life], *Revue de la Société Scientifique d'Hygiène Alimentaire et de l'Alimentation Rationnelle de l'Homme*, 1905.

Javal, A.L. *Obésité, Hygiene et Traitement* [Obesity, Hygiene and Treatment], unpublished thesis, Paris, 1900.

Labbé, H. *Hygiène Sociale. Enquête sur l'Alimentation d'une Centaine d'Ouvriers et d'Employés Parisiens* [Social Hygiene: Food Survey of a Hundred Workers and Parisian Employees], Enquête Présentée à la IVe Section du Congrès International de la Tuberculose, 2–7 Octobre 1905, Paris, 1905.

Leven, G. *De l'Obésité* [On Obesity], Paris, 1901.

Lhuissier, A. *Alimentation Populaire et Réforme Sociale* [Popular Food and Social Reform], Paris, 2007.

Pascault, L. 'L'Arthritisme, Maladie de Civilisation' [Arthritis: a Disease of Civilization], *Revue des Idées*, 25, 1906.

Poulain, J.P. *Manger Aujourd'hui. Attitudes, Normes, Pratiques* [Eating Today: Attitudes, Norms, Practices], Toulouse, 2002.

Quetelet, A. *Anthropométrie ou Mesure des Différentes Facultés de l'Homme* [Anthropometry or Measuring the Various Faculties of Man], Brussels, 1870.

Raffray, A. *Le Péril Alimentaire* [The Danger of Food], Paris, 1912.

Ramond, F. and Oulmont, P. *L'Obésité, Symptomatologie et Étiologie, Anatomie et Physiologie Pathologique, Traitement* [Obesity: Etiology and Symptomatology, Pathological, Anatomy and Physiology, Treatment], Paris, 1907.

Reboux, P. *Nouveaux Régimes ou l'Art d'Accommoder Selon la Gastronomie les Ordonnances des Médecins* [New Schemes or the Art of Cooking Modified According to the Ideas of Doctors], Paris, 1931.

Thoms, U. 'Des Perceptions de la Minceur et de l'Obésité de 1850 à nos Jours' [Perceptions of Slimness and Obesity from 1850 to the Present], in Audouin-Rouzeau, F. and Sabban, F. (eds), *Un Aliment Sain dans un Corps Sain. Perspectives Historiques*, Tours, 2007, 319–36.

Worthington, L.S. *De l'Obésité: Etiologie, Thérapeutique et Hygiène* [On Obesity: Etiology, Therapy and Hygiene], unpublished medical thesis, Paris, 1877.

Chapter 13

Slimming Through the Depression: Obesity and Reducing in Interwar Britain

Ina Zweiniger-Bargielowska

In December 1929 the obesity expert W.F. Christie declared that Britain was in the midst of a 'slimming' craze and in 1933 the *Lancet* noted that in 'these days of "slimming"' there was no more popular subject of discussion among the laity than the reduction in weight.[1] Doctors' growing interest in obesity and the rise of a flourishing self-help reducing literature may seem incongruous for an era dominated by economic depression, unemployment, poverty and hunger marches. Concern about weight-gain did not eclipse the 'Hungry England' debate of the 1930s, but evidence such as dietary surveys, medical debates and self-help literature makes it possible to construct a more complex pattern of eating habits which posits interwar Britain at a transitional stage in European food systems. A study of obesity, therefore, provides an important supplement to the historiography of diet, nutrition and public health in interwar Britain, which is dominated by a focus on hunger and undernutrition.[2]

Obesity was determined not only by class, but also by age and gender. Not surprisingly, the condition was prevalent within the middle class and associated with the growing comforts of middle age. This chapter explores representations of obesity and juxtaposes the common emphasis on female reducing with the image of the overweight middle-aged businessman. Described by Christie as a 'deviation from the normal', the obese were subjected to a moralizing discourse which condemned women's excessive slimming in the pursuit of fashion and castigated obese men's self-indulgence as subverting dominant codes of masculine self-restraint.[3] As Leonard Williams, who emphasized the essential difference between men and women, put it, 'No man has any right to be really fat; no woman has any right to be really thin'.[4] Reducing systems ranged from drug treatments, which were criticized as unsatisfactory and even dangerous, to the age-old combination of dietary restraint and exercise, which emphasized conduct rather than calories.

---

1 *British Medical Journal* 28 December 1929: 1198, *Lancet* 14 October 1933: 871, Graves and Hodge 1940: 189–90, 230–31.

2 Webster 1983, Burnett 1979, Jones 1994, Oddy 2003, Vernon 2007.

3 *British Medical Journal* 28 December 1929: 1198.

4 Williams 1926: 125.

Excess consumption among the wealthy was not a new phenomenon but the imperative of restraint acquired a greater prominence in response to rising living standards and urbanization from the late nineteenth century. This was accompanied by persistent income inequality and working-class poverty stood in marked contrast with the emergence of affluence among the middle class, particularly during the 1930s. A reducing culture was part of a wider life reform movement which advocated a regimen of bodily discipline encompassing dietary restrictions, exercise or physical culture, sun-bathing and dress reform to counter the negative impact of modern urban sedentary lifestyles. These developments coincided with the proliferation of visual images of beautiful bodies and the rise of eugenicist and social Darwinist fears about racial deterioration. Public health advocates and doctors focused on improving the health of the poor, but this was accompanied by growing awareness of the need to manage the bodies of the wealthy.[5] The promotion of a regimen to raise the standard of racial health acquired urgency in the wake of the dysgenic disaster of the First World War and Sir George Newman, Chief Medical Officer at the newly established Ministry of Health, aimed to educate the public by preaching hygienic habits and self-control. This emphasis on the cultivation of health as a personal responsibility and civic duty was echoed by pressure groups such as Sir William Arbuthnot Lane's New Health Society, launched in 1925. The society's members included Leonard Williams, Fredrick Hornibrook, author of the best-selling *The Culture of the Abdomen: The Cure of Obesity and Constipation*, and Hornibrook's wife, Ettie Rout, whose publications addressed the specific needs of women.[6]

The statistical evidence is limited, but dietary surveys show that extensive undernutrition among the poor coincided with excess consumption among the higher income groups during the interwar years. Research was focused on the working class, but a handful of surveys covering the entire social spectrum demonstrate a correlation between class and calorie consumption. This increased with rising incomes and was highest among the wealthiest. A 1931 Medical Research Council survey of St Andrews, which aimed to establish quantitative data on the 'normal diets' of families, observed this differential of food expenditure and energy intakes. Daily energy intakes per man value ranged from 3,638 kilocalories (kcals) among professionals and 3,333 kcals among sedentary workers to 3,095 kcals among manual workers and 2,089 kcals among the unemployed.[7] John Boyd Orr's influential *Food, Health and Income* highlighted differences in food expenditure and energy intakes with 3,326 and 2,317 kcals per day respectively among the top and bottom 10 per cent in terms of income.[8] A survey by Crawford and Broadley

---

5   Forth and Carden-Coyne 2005, Schwarz 1986, Gilman 2004. For a discussion of similar themes in the German context, see Merta 2000.

6   Hornibrook 1924, Rout [Hornibrook] 1925, 1934, Foucault 1978, Zweiniger-Bargielowska 2005, Zweiniger-Bargielowska 2007, Jones 1986, Whorton 2000.

7   Medical Research Council 1931, hereafter MRC 151: 23, 27.

8   Orr 1936, 24, 34.

in the late 1930s noted a similar differential daily energy intakes ranging from 3,536 kcals among the very rich to 2,335 among the poorer sections of the working class.[9] While these data have been used extensively to highlight dietary deficiencies among the working class, my purpose is to draw attention to excess consumption amongst the middle classes. With daily male energy requirements estimated at about 3,000 kcals, ranging from 2,700 kcals for sedentary workers to 4,500 kcals for men engaged in heavy manual labour, these surveys suggest that the food consumption of middle-class men was consistently above recommended levels.[10] Struck by the inverse correlation between calorie consumption and energy requirements, the St Andrews study accepted that the sedentary worker was 'probably overfed', but did not necessarily consider the energy intake as excessive and claimed that obesity was 'relatively rare'. The study concluded that the most plausible explanation of the disproportional energy intake between sedentary workers and the professional middle class was their better physique and habit of more strenuous exercise by contrast with manual workers.[11] There are no obesity statistics for this period, but it is worth noting that rates of coronary heart disease among men began to rise in the 1920s and excess mortality in Social Class I during the 1930s was attributed to degenerative diseases of affluence.[12] This included obesity, which became a recurrent topic in British medical journals during the 1930s, suggesting a growing concern among doctors about the waistlines of their middle-class patients.

By contrast with late twentieth-century data, which show an inverse correlation between socioeconomic status and obesity, the interwar evidence points towards a positive correlation.[13] Economic constraints prevented excess consumption among wide sections of the working class, many of whom suffered from undernutrition, and obesity was not perceived to be common among the upper class, whose lifestyle included the pursuit of vigorous outdoor activities. Rather, the condition was associated with the petty comforts of sedentary suburban middle-class life or the consumerism of new wealth. Leonard Williams contrasted the leanness of the typical aristocrat 'whose ancestors through the ages have had no necessity for hoarding fat' with 'the profiteer, and the *nouveau riche*...always depicted as bull-necked and pot-bellied – which he generally is'.[14]

Obesity was associated with middle age.[15] Williams deplored middle-age spread, arguing that the 'ordinary man begins to "rot" at 30 years of age, and when he reaches 50 the sinister process is in full swing' and he blamed wives who overfed their husbands because a fat man was 'easy-going, yielding, uncritical; stupid in

---

9 Crawford and Broadley 1938: 154. Similar disparities were also observed by Rowntree at the turn of the century, Rowntree 1901.
10 Orr 1936: 32–3.
11 MRC 151: 22, 24–6.
12 Halsey and Webb 2000: 101–3, 110–13.
13 Offer 2006: 152–3.
14 Williams 1926: 22.
15 Christie 1927: 34, *British Medical Journal* 15 August 1936: 344.

fact'.[16] A drawing of an obese middle-aged man longingly admiring a fashionably slim young woman illustrates Christie's chapter entitled 'Self-Indulgent Fat or the Unfashionable Figure'.[17] This character was also a stock figure in adverts for remedial products such as abdominal belts. One example depicts a thin man and another with a protruding abdomen and asked:

> Young Man which will be you at 40? The 'middle-aged spread' is a real menace that our sedentary modern life forces on us…[T]he 'Danger Curve' destroys vigour and invites serious trouble…Linia Shorts are the modern answer…they support the abdominal muscles and keep them healthy by continuous unobtrusive massage.[18]

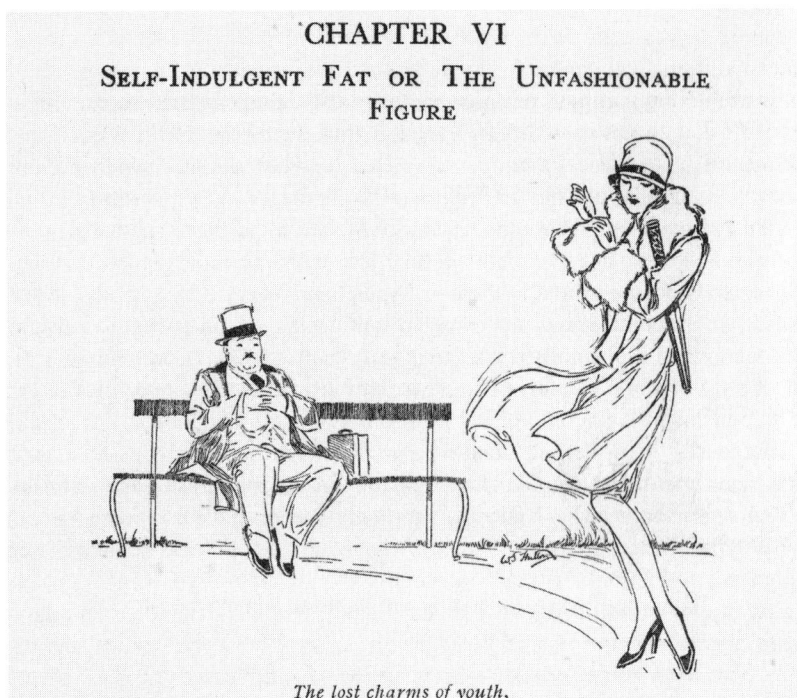

Figure 13.1 'The Lost Charms of Youth'

Source: Christie 1927: 32.

---

16   Williams 1925: 7, Williams 1926: 3. The middle-aged, middle-class man was also the focus of Hornibrook 1924, 1927.

17   Christie, *Surplus Fat*: 32.

18   *Men Only*, May 1936: 145. These products were widely advertised also in *New Health*, for example June 1934, *Picture Post*, for example 25 February, 1939. *New Health* was published by the New Health Society.

The association of middle-age spread with complacency and self-indulgence among the middle class was a new phenomenon of the twentieth century.[19] George Orwell, commenting on the portrayal of the middle aged as obese in contrast with youthful slim sex appeal in Donald McGill's comic postcards, argued that the aspiration of a youthful appearance among middle-class middle-aged people illustrated an important shift in attitudes:

> to look young after, say, thirty is largely a matter of wanting to do so…The impulse to cling to youth at all costs, to attempt to preserve your sexual attraction, to see even in middle age a future for yourself and not merely for your children, is a thing of recent growth.[20]

The middle-aged, middle-class businessman as a sedentary worker emerges as the principal sufferer from obesity, but of course the condition was not confined to men and excess weight also occurred among women. As Christie put it, obesity was 'probably more common in the circles of the well-to-do, and more common still in those with nothing-to-do' such as married middle-class women. Women were particularly prone to weight gain after child birth when many new mothers forsook athletic pursuits and acquired the habit of 'highly nutritive foods … to maintain their own strength, and the better to suckle their infant'. Another period of weight-gain was during menopause when women commonly became less active but often developed a habit of eating more.[21] Christie's subsequent analysis of 184 cases, of whom about three-quarters were female, highlighted gender differences after the onset of obesity. The condition often started during puberty among girls and continued to be associated with women's reproductive function by contrast with a typical male pattern of onset during middle age.[22] Evidence of gender differences in food consumption suggests that women were less likely to be obese than men. Derek Oddy concludes that before the First World War whereas men 'ate and drank to the point of becoming stout', women 'through under-eating, inclined to illness'. Working-class married women's self-sacrifice to ensure the breadwinner's health and happiness is widely documented.[23] Given the strain on many working-class budgets, this pattern persisted during the interwar years.[24] Female dietary restraint was not just due to poverty. Moderate eating signified control over appetite and represented politeness and refinement. Notions of excess weight are culturally constructed and the celebration of a slim ideal in women's fashion in the 1920s gave rise to a female reducing culture which many doctors condemned as excessive. Slimming was thus motivated not only by excess weight,

---

19   Benson 1997: 17.
20   Orwell 1946: 91, 94–5.
21   Christie 1927: 34–5.
22   Christie 1937: 60.
23   Oddy 2003: 70, Rowntree 1901: 135, Ross 1993.
24   Spring Rice 1939.

but also by new standards of how much was too much and the desire to maintain a youthful body, which became embedded in middle-class culture during the interwar years.

Doctors attributed obesity primarily to exogenous or alimentary factors, that is excessive food consumption, although a small minority of cases was perceived as endogenous, caused by endocrine or glandular malfunctioning. Obesity was held responsible not only for ill-health and premature death, but was also portrayed as ugly and morally repugnant. This was not merely an individual, private matter and for Christie the 'maintenance of one's body in its most agreeable and effective form [was] the obvious duty of every member of a civilized community'. Therefore:

> surplus fat should neither be tolerated with resignation, nor left for concealment to the tailoring craft. The social unhappiness, physical inefficiency and shortened tenure of life which an excessive burden entails, are sufficient reasons.[25]

Williams was even more vehement in his moral condemnation of fat which he considered not only as contemptible and disgusting, but also pathological and degenerate. Obesity was due to self-indulgence and greed and he claimed that the condition was common among criminals, embezzlers and homosexuals.[26] The association of obesity with financial crime was rejected by other doctors, such as Donald Hunter, who nevertheless acknowledged that many obese patients were lying or deceiving themselves with regard to their eating habits and that the first step was to persuade them that 'weight *can* be reduced by diet' from which there must be no departure.[27]

Maintaining that moderation in diet was the 'surest method of qualifying for longevity', Williams claimed that more people 'floated into their coffins on a flood of beef-tea and milk than ever arrive there by the ravages of disease'.[28] The long-standing perception that fat men, in particular, died young was confirmed by statistics of the high mortality risk of obesity among middle-aged men compiled by American life insurance companies.[29] Emphasizing the link between obesity and cardiovascular, renal and gastrointestinal diseases, Williams portrayed fat as a 'malignant and merciless parasite' and he concluded that obesity constituted a serious menace to the health and efficiency of the individual.[30] These arguments were endorsed by Ernest Bulmer, whose discussion warned that if the national tendency to be overweight were to 'continue to grow unchecked, the mortality

---

25   Christie 1930: 817, Christie 1927: v.
26   Williams 1926: 1, 29, 72.
27   *Lancet* 28 October 1933: 994–5 [emphasis in original].
28   Williams 1925: 9–10.
29   Fisher and Fisk 1915, Whorton 2000: 191–6, 200.
30   Williams 1926: 37, 59, see 36–59 for a discussion of the various diseases linked with obesity.

from the degenerative, non-bacterial diseases will diminish the average expectation of life'.[31]

Obesity conflicted with the dominant ideal of physical beauty based on a classical aesthetic model, and images of scantily-clad beautiful male and female bodies were celebrated in popular culture during the 1930s. Paying homage to the Greek ideal, Hornibrook maintained that there was 'not one solitary instance of a beautiful form fashioned in fatness'. He was disgusted by the common sight of 'fat, ugly, clumsy bodies' and pitied the obese with their protuberant belly and ponderous buttocks which handicapped 'their cumbrous waddle through life'.[32] It would be wrong to assume that men did not care about their appearance. Recent research demonstrates that men paid close attention to fashion and remedial products intended to improve the sagging, flabby and balding middle-aged male body were widely advertised in the interwar years. These features can be juxtaposed with the beautiful male physique, worshipped by the physical culture movement, which originated in the late nineteenth century and was re-launched after the First World War. The obese were targeted by leading physical culture promoters such as Eugen Sandow, Thomas Inch and Jørgen Peter Müller, who saw exercise as crucial in men's quest for a taut and muscular body.[33]

The discourse of beauty was even more significant for women and the interwar years witnessed not only a major transformation in women's fashion but also the rise of a commercial cosmetics and beauty culture. The new fashions required a boyish, slender figure which contrasted not only with the heavily corseted hour-glass silhouette of the late Victorian period, but also the Venus de Milo counterpart of the male Greek beauty ideal of the early twentieth century.[34] United in their condemnation of the corset, physicians and physical culture advocates extolled the outlines of the Venus whose straight waist, well-developed oblique muscles and broad hips represented the healthy, efficient physique of the 'race mother'.[35] Hornibrook who, in the early 1900s, had praised Ettie Rout as the 'only woman he had ever met who came up to the Venus de Milo standard', was part of a chorus of commentators who deplored women's slimming as excessive in the interwar years.[36] Accompanied by an illustration showing two slim flappers looking at the Venus de Milo, who did not 'see what there [was] to admire', an article by Hornibrook in *New Health* condemned the craze for slimness as a real danger that increased the risk of tuberculosis and turned women into 'nervous wrecks. Excessive dieting not only resulted in a loss of health, but also an inevitable loss of beauty by making women

---

31 *British Medical Journal* 4 June 1932: 1024.
32 Hornibrook 1924: 17, Hornibrook 1927: 88.
33 Horwood 2005, Greenfield, O'Connell and Reid 1999, Zweiniger-Bargielowska 2005, 2006, Mosse 1996, Bourke 1996.
34 Zweiniger-Bargielowska 2001: 183–97, Bingham 2004, Getsy 2004, Horwood 2005.
35 Müller 1911, Arbuthnot Lane 1909, Davin 1978.
36 Tollerton 2001: 75.

First Flapper : " I should just hate to have a figure like that !
  Wouldn't you ? "
Second Flapper : " I really don't see what there is to admire in it ! "

Figure 13.2 Flappers' reactions to the Venus di Milo

Source: *New Health* August 1929: 30.

thin and scraggy with an 'aged, deeply-lined and discontented appearance'.[37] Williams yet again raised the spectre of sexual deviance, claiming that the new female fashion of extreme slimness was not dictated by a desire to attract normal manly men, but rather a 'ruse to find favour in the eyes of the degenerates and homosexuals'.[38] Summing up as harmful the physical and psychological effects of slimming – such as lowered resistance to tuberculosis, an irritable disposition, a haggard, drawn expression which even 'liberal make-up will not successfully hide' and the danger of anorexia nervosa, which commonly originated with a girl being teased because she was getting plump – the *British Medical Journal* lamented that 'the sex which for many years injured its health by tight lacing is not likely to be deterred from slimming by such considerations. The dictates of fashion will be paramount'.[39] Lane was somewhat more sympathetic and welcomed women's endeavour to avoid the ever-present danger of putting on flesh which he contrasted with the pitiful condition of middle-aged men. Nevertheless, he upheld the Venus de Milo as the feminine ideal and castigated women who carried slimming too far because a 'vigorous, well-developed physique' was vital in view of women's responsibility as 'race mothers'.[40]

This vilification of middle-aged men's self-indulgent greed, and women's equally selfish pursuit of fashion at the expense of health, parallels the long-established moral condemnation of the feckless poor. All fell short of their civic obligation to cultivate health, which was imperative for Britain's future as a leading imperial power. This argument is most prominently associated with Sir George Newman, who downplayed the significance of lack of money and, instead, emphasized the importance of ignorance and bad habits as the leading cause of ill-health and physical degeneration before 1914. Claiming that ignorance and lack of self-control were the two roots of evil, Newman argued in 1907 that the most urgently needed public health reform was a reform of personal life. Advocating moderation and cleanliness in the ways and habits of life, the maintenance of a high standard of personal health was not only the best preventive of disease, but also 'Every man's first contribution to the State' because the physical health and fitness of the people was 'the primary asset of the British Empire'.[41] Newman restated his position after the war in his annual reports *On the State of Public Health*, which drew attention to 'our faulty habits and customs in respect of dietary' and resulted in his view in a substantial degree of impairment of physique. In 1926, he held excessive and unsuitable food, combined with lack of fresh air and exercise, responsible for sowing the seeds of degeneration. He accepted that some persons no doubt were under-fed in 1931, but nevertheless argued that many were 'over-

---

37 *New Health* August 1929: 30–31.
38 Williams 1926: 77.
39 *British Medical Journal* 27 November 1937: 1077.
40 *New Health* January 1931: 13, Lane 1936: 175–6, 185-6, Lane 1930: 865–8.
41 Newman 1907: vi, 191, 193–4, Smith and Nicolson 1995, Welshman 1997.

fed – giving their poor bodies little rest, clogging them with yet more food'.[42] Newman saw health as a personal responsibility and he advocated health education to instil a health conscience which depended on the 'enlightened goodwill of the individual citizen'.[43] Newman was even more emphatic in his report of 1929 when he declared that:

> *Health can only be achieved by the people themselves*...a nation becomes physically strong and healthy if each individual so cultivates his own body and mind as to live at the top of his physical, mental and moral capacity. This means an ordered way of life...every adult must discipline and train himself, or be trained,...to understand and practise this art of Living – much of the essence of which is contained in the Greek aphorism, 'know thyself and be moderate in all things'.[44]

Reducing systems included drugs, dietary restraint and exercise. Thyroid extract, widely prescribed in the early twentieth century, was condemned in the mid-1920s and by the 1930s generally rejected.[45] New metabolic drugs such as dinitrophenol were briefly popular but discredited following the death of a female slimmer. Maintaining that it was possible to discuss slimming agents in a 'note on cosmetics', Sir Arthur MacNalty, Newman's successor as Chief Medical Officer, advised against slimming drugs or preparations except on medical advice and declared that dinitrophenol should not be taken under any circumstances.[46] This illustrated the consequences of women's excessive pursuit of slenderness and Christie summed up the dangers of slimming, which were first, irrational diets such as semi-starvation or unbalanced meals, second, unsuitable cases, including those already thin enough; and, finally, the use of drugs, which was in the great majority of cases unnecessary. In subsequent correspondence, Christie again rejected drugs because any short-cut to slimness was 'fraught with perils', for which there was no justification when the cure of obesity by correct dieting was safe and certain.[47]

Dietary restrictions were central, but in contrast with the late twentieth-century reducing culture, calories did not occupy a prominent position in many interwar reducing manuals. Instead, advice focused on conduct and manuals portrayed the

---

42   Ministry of Health, *On the State of Public Health: Annual Report of the Chief Medical Officer of the Ministry of Health for the Year 1921*, London: 1922: 86 [hereafter MH, *Annual Report*], MH, *Annual report 1926,* 266, MH, *Annual Report 1930*: 162. See also Newman 1939: 347–52.

43   MH, *Annual Report 1922*: 171.

44   MH, *Annual Report 1929*: 206, 207–8 [emphasis in original].

45   *Lancet* 3 January 1925: 36, *Lancet* 24 March 1934: 656, *British Medical Journal* 21 April 1934: 700.

46   *Lancet* 24 March, 7 April 1934: 652, 746, MH, *Annual Report 1936*: 187, *British Medical Journal* 27 November 1937: 1076–7.

47   *Lancet* 24 March 1934: 656, 1 June 1935: 1301.

practice of a regimen of bodily discipline as central to the cultivation of selfhood, which was a public duty of citizenship. Echoing Newman's call for self-discipline, Leonard Williams scorned people who sought quick relief from quack remedies instead of accepting the necessity of a slow and disagreeable operation of diet and regime. Successful reducing demanded tireless watchfulness and should begin with three days of fasting, followed by three days of light eating. Williams did not even mention calories and prescribed either a meat-rich, low-carbohydrate diet or an 'unfired' diet based on dairy products, vegetables, fruit and wholemeal bread. This was to be combined with exercise and Williams recommended Hornibrook's system of abdominal exercise.[48] Conduct was central to Williams's *Middle Age and Old Age*, which urged the necessity of reforming the present habits, dietetic and otherwise as the best means of arriving at Old Age. Similarly, the *Science and Art of Living* made the case for a simpler life to improve health, efficiency and longevity because humans were subject to physiological laws which were violated above all by gluttony, the crime of civilization.[49]

Calories were not entirely absent from British reducing manuals and Christie, who advocated a gradual weight reduction of one or two pounds per week, suggested consuming between 500 and 1,000 kcals less than required in a maintenance diet. Menus were to be planned using a skeleton diet of lean meat, fruit, vegetables and skimmed milk. Additional food, up to 1,200 kcals per day, should accord with the patient's desires. In view of the common occurrence of 'breaks away' from dieting due to weak wills, Christie also proposed a slowly reducing diet of 1,325 kcals per day because the 'enthusiastic co-operation of the patient is essential if success is to be achieved'. Dietary restraint should be combined with exercise and Christie also recommended Hornibrook's system as well as pedestrianism, a walking cure based on increasingly lengthy and arduous walks.[50] Another example, Claxton's *Weight Reduction: Diet and Dishes*, the most calorie-conscious British manual, provided a range of options from a starvation diet to qualitative restriction for 'milder' cases. Positively reviewed in the *Lancet*, the book's main attraction was a large number of recipes, which showed that a slimming diet need not be monotonous and unpalatable.[51]

Aimed above all at middle-aged businessmen, who were urged to cultivate 'those parts of the body that, owing to man's posture and to his civilized habits are the most neglected…viz., the muscles of the abdomen and the organs of digestion and excretion', Hornibrook's *Culture of the Abdomen* did not mention calories. Influenced by Lane's emphasis on the dangers of constipation, Hornibrook stressed the link between a bloated abdomen, constipation and obesity. He sought to 'localize effort to the abdominal region, to promote more abdominal activity, and to concentrate on correct posture. Widely recommended by physicians,

---

48 Williams 1926: 122–4, 135, 140–47, 162–5
49 Williams 1925: iv, Williams 1924: 7–9.
50 Christie 1927: 66–80, Christie 1937: 77–91, 179–80.
51 Claxton 1937: vii, 41–7, 51–9, 62, *Lancet* 16 October 1937: 910.

Hornibrook's system consisted of seven minutes daily practice of abdominal exercises based on native dances, coupled with regular retraction of the abdominal muscles which should be practised throughout the day. These exercises, combined with thorough mastication, liberal quantities of water and 'proper habits of eating' guided by the golden rule of moderation, activated the bowel and reduced weight in the abdominal area.[52] Adapting Hornibrook's system to women, Rout highlighted the link between digestive organs and women's reproductive function. In *Sex and Exercise*, reissued in 1934 as *Stand Up and Slim Down: Being Restoration Exercises for Women with a Chapter on Food Selection in Constipation and Obesity*, she claimed that abdominal, pelvic and hip exercises combined with a high fibre diet were essential to acquire a slim and beautiful figure. More importantly, these practices enhanced women's efficiency with regard to sexual intercourse, pregnancy and labour, imperative in view of women's role as 'race mothers'.[53] The aim of 'Racial Health' was similarly advanced by the most famous female physical culture organization of the period, the Women's League of Health and Beauty launched in 1930.[54]

The concurrence of obesity with extensive undernutrition exemplifies the social and economic inequality of interwar Britain, characterized by the emergence of affluence within the middle classes and persistent poverty among sections of the working class. The disparate food systems of the middle and working classes represented opposing poles and yet, dietary patterns and habits of both groups were also perceived to exhibit common elements because fat, emaciated and stunted bodies deviated from conceptions of the normal. All were subjected to a moralizing discourse which saw the cultivation of health as a personal responsibility within reach of all who practiced a regimen. As *New Health* put it,

> Let the public remember that poor physique, ill-health, and…premature disease are all avoidable if the people obey the simple laws of health, and that no one has any right to inflict their unhappiness, their disability to work and other associated evils upon the members of the community. Disease in most forms is essentially criminal, since its avoidance is in the hands of the people themselves.[55]

This perspective was echoed by obesity experts and Ministry of Health officials who vilified not only the obese and excessive slimmers, but also the feckless poor because they violated the golden rule of moderation, whether through greed, vanity or ignorance and thereby fell short in their civic duty to cultivate their health for the sake of the nation and empire.

---

52   Hornibrook 1924: 3–5, 38, 45–7. Lane wrote the preface and endorsed Hornibrook's system in *New Health* November 1927: 17–18, see also *British Medical Journal* 15 August 1936: 344, 27 November 1937: 1071–2.

53   Rout [Hornibrook] 1925, 1934.

54   Matthews 1990: 25.

55   *New Health* January 1927: 6.

A reducing culture flourished until the outbreak of war but, following the introduction of food rationing in 1940, doctors lost interest in obesity. No reducing manuals were published in Britain between 1940 and the early 1950s and slimming products vanished during the era of austerity.[56] This evidence draws attention to the limits of personal responsibility and highlights the significance of a heavily regulated, restrictive food regime combined with more active lifestyles during the 1940s. A predominantly female reducing culture reappeared during the late 1950s and in subsequent decades an ever expanding commercial slimming culture coincided with substantial weight-gain and rising obesity.[57]

References

Benson, J. *Prime Time: A History of the Middle Aged in Twentieth-Century Britain*, London, 1997.

Bingham, A. *Gender, Modernity and the Popular Press in Interwar Britain*, Oxford, 2004.

Bourke, J. *Dismembering the Male: Men's Bodies, Britain and the Great War*, Chicago, 1996.

Burnett, J. *Plenty and Want. A Social History of Diet in England from 1815 to the Present Day*, 2nd ed., London, 1979.

Christie, W.F. *Surplus Fat and How to Reduce It*, London, 1927.

Christie, W.F. 'Corpulence', in Arbuthnot Lane, Sir W. (ed.), *The Golden Health Library: A Complete Guide to Golden Health for Men and Women of all Ages*, vol. 3, London, 1930.

Christie, W.F. *Obesity: A Practical Handbook for Physicians*, London, 1937.

Claxton, E.E. *Weight Reduction: Diet and Dishes*, London, 1937.

Crawford, Sir W. and Broadley, H. *The People's Food*, London, 1938.

Davin, A. 'Imperialism and Motherhood', *History Workshop* 5, 1978, 9–65.

Department of Health. *Obesity: Reversing the Increasing Problem of Obesity in England*, London, 1995.

Fisher, I. and Fisk, E.L. *How to Live: Rules for Healthful Living Based on Modern Science*, New York, 1915.

Forth, C.E. and Carden-Coyne, A. (eds), *Cultures of the Abdomen: Diet, Digestion and Fat in the Modern World*, Basingstoke, 2005.

Foucault, M. *Discipline and Punish: The Birth of the Prison*, New York, 1978.

Getsy, D.J. *Body Doubles: Sculpture in Britain, 1877–1905*, New Haven, CT, 2004.

Gilman, S. *Fat Boys: A Slim Book*, Lincoln, NE, 2004.

---

56 Zweiniger-Bargielowska 2000.

57 Yudkin 1958, Offer 2006: 138–69, Department of Health 1995, *Social Trends* 29, 1999: 122, *Social Trends* 33, 2003: 140.

Graves, R. and Hodge, A. *Long Week-end: A Social History of Great Britain, 1918–1939*, London, 1940.

Greenfield, J., O'Connell, S. and Reid, C. 'Gender, Consumer Culture and the Middle-Class Male', in Kidd, A. and Nichols, D. (eds), *Gender, Civic Culture and Consumerism: Middle-Class Identity in Britain, 1800–1940*, Manchester, 1999.

Halsey, A.H. and Webb, J. (eds), *Twentieth-Century British Social Trends*, Basingstoke, 2000.

Hornibrook, F.A. *The Culture of the Abdomen: The Cure of Obesity and Constipation*, London, 1924.

Hornibrook, F.A. *Physical Fitness in Middle Life*, London, 1927.

Horwood, C. *Keeping Up Appearances: Fashion and Class between the Wars*, Stroud, 2005.

Jones, G. *Social Hygiene in Twentieth-Century Britain*, London, 1986.

Jones, H. *Health and Society in Twentieth-Century Britain*, London, 1994.

Lane, W.A. 'Civilization in Relation to Abdominal Viscera, with Remarks on the Corset', *Lancet*, 13 November 1909, 1416–8.

Lane, Sir W.A. 'Women and the Race', in Lane, 1936.

Lane, Sir W.A. *Every Woman's Book of Health and Beauty*, London, 1936.

Matthews, J.J. 'They Had Such a Lot of Fun: the Women's League of Health and Beauty between the Wars', *History Workshop* 30, 1990, 22–54.

Medical Research Council, *A Study of Nutrition: an Inquiry into the Diet of 154 Families of St Andrews, Special Reports Series No 151* [E.P. Cathcart and A.M.T. Murray], London, 1931.

Merta, S. '"Keep Fit and Slim!" Alternative Ways of Nutrition as Aspects of the German Health Movement, 1880–1930', in ICREFH V, 170–202.

Mosse, G.L. *The Image of Man: the Creation of Modern Masculinity*, New York, 1996.

Müller, J.P. *My System for Ladies*, London, 1911.

Newman, G. *The Health of the State*, London, 1907.

Newman, Sir G. *The Building of a Nation's Health*, London, 1939.

Oddy, D.J. *From Plain Fare to Fusion Food: British Diet from the 1890s to the 1990s*, Woodbridge, 2003.

Offer, A. *The Challenge of Affluence: Self-Control and Well-Being in the United States and Britain since 1950*, Oxford, 2006.

Orr, J.B. *Food, Health and Income: A Report on a Survey of Adequacy of Diet in Relation to Income*, London, 1936.

Orwell, G. 'The Art of Donald McGill', in Orwell, G., *Critical Essays*, London, 1946.

Ross, E. *Love and Toil: Motherhood in Outcast London, 1870–1918*, New York, 1993.

Rout [Hornibrook], E.A. *Sex and Exercise: A Study of the Sex Function in Women and its Relation to Exercise*, London, 1925.

Rout [Hornibrook], E.A., *Stand Up and Slim Down: Being Restoration Exercises for Women with a Chapter on Food Selection in Constipation and Obesity*, London, 1934.

Rowntree, B.S. *Poverty: A Study of Town Life*, London, 1901.

Schwarz, H. *Never Satisfied: a Cultural History of Diet, Fantasies and Fat*, New York, 1986.

Smith, D. and Nicolson, M. 'Nutrition, Education, Ignorance and Income: A Twentieth-Century Debate', in Kamminga, H. and Cunningham, A. (eds), *The Science and Culture of Nutrition, 1840–1940*, Amsterdam, 1995, 288–318.

Spring Rice, M. *Working-Class Wives: Their Health and Conditions*, Harmondsworth, 1939.

Tollerton, J. 'A Lifetime of Campaigning: Ettie Rout, Emancipationist beyond the Pale', in Mangan, J.A. and Hong, F. (eds), *Freeing the Female Body: Inspirational Icons*, London, 2001.

Vernon, J. *Hunger: A Modern History*, Cambridge, MA, 2007.

Webster, C. 'Healthy or Hungry 30s', *History Workshop* Journal 13, 1983, 11029.

Welshman, J. '"Bringing Beauty and Brightness to the Back Streets": Health Education and Public Health in England and Wales', *Health Education Journal* 56, 1997, 199–209.

Whorton, J.C. *Inner Hygiene: Constipation and the Pursuit of Health in Modern Society*, New York, 2000.

Williams, L. *The Science and Art of Living*, London, 1924.

Williams, L. *Middle Age and Old Age*, London, 1925.

Williams, L. *Obesity*, London, 1926.

Yudkin, J. *This Slimming Business*, London, 1958.

Zweiniger-Bargielowska, I. '"The Culture of the Abdomen": Obesity and Reducing in Britain, c.1900–1939', *Journal of British Studies* 44, 2005, 239–73.

Zweiniger-Bargielowska, I. 'Building a British Superman: Physical Culture in Interwar Britain', *Journal of Contemporary History* 41, 2006, 595–610.

Zweiniger-Bargielowska, I. 'Raising a Nation of "Good Animals": The New Health Society and Health Education Campaigns in Interwar Britain', *Social History of Medicine* 20, 2007, 73–89.

Zweiniger-Bargielowska, I. 'The Body and Consumer Culture', in Zweiniger-Bargielowska, I. (ed.), *Women in Twentieth-century Britain*, Harlow, 2001.

Zweiniger-Bargielowska, I. *Austerity in Britain: Rationing, Controls and Consumption, 1939–1955*, Oxford, 2000.

Chapter 14

Socialism and the Overweight Nation: Questions of Ideology, Science and Obesity in Czechoslovakia, 1950–70

Martin Franc

At the beginning of the 1950s obesity was a marginal problem in Czechoslovakia.[1] Food was still rationed, which definitely did not support overeating. Although at the end of the 1940s there was rapid growth in the energy value of the diet, economic troubles at the beginning of the 1950s resulted in a renewed decline. The situation was partly improved by the parallel free market, where high prices prevailed. However, the overall atmosphere in the first half of the 1950s with regard to fatness is interestingly described by an ordinary woman who permanently struggled with excess weight:

> Let's put aside the political monstrosity of those years, but there is something we cannot quite deny: a certain truthfulness in the faith that everyone is equal and that bodily features, features of appearance are secondary in importance.[2]

Nevertheless, the situation was not ideal, as the vulgar popular jibes against Marta – the corpulent wife of President Klement Gottwald – demonstrated.[3] In the popular literature, her figure seemed to symbolize caricatures, like those of the nineteenth century, which typified townspeople as having a big belly. Not all negative figures, however, were represented as overweight people in aggressive

---

1   The term obesity seems to have had a much more general meaning in Czechoslovakia in the 1950s and 1960s than would be the case today. Czechoslovakia, like much of central Europe, used the Broca Index, which allowed weights for women to vary by plus or minus 15 per cent and for men by plus or minus 10 per cent. Terms like *otylost* (excess weight) and *obezita* (obesity) were used interchangeably in Czechoslovakia for people who were more than 5 per cent above Broca's 'ideal weight'. In 1963, Czechoslovak research suggested that 'heavy' obesity, i.e. those adults weighing more than 20 per cent above 'ideal weight', affected 7.6 per cent of men and 22.4 per cent of women.

2   Štolbová 2000: 7.

3   Marta Gottwaldová (1899–1953), wife of Klement Gottwald from 1928. In addition to being obese, she was also famous by imitating the lifestyle of the old bourgeoisie. The memoirs of her doctor, the famous physician, Josef Charvát, indicate that her obesity might have been caused by a thyroid disorder. For jibes see Rogl 2005.

satirical pictures. The complete opposite was quite frequently used, that is, extremely skinny figures, such as America's Uncle Sam or the West German Chancellor Konrad Adenauer; even cartoons depicting young 'fops' and elegantly dressed ladies from bourgeois circles were usually depicted as very slim.[4] The statuesque figures of labourers formed a counterpart to both overweight and unnaturally slim people. Workers included women, because emancipation in Czechoslovakia during the first half of the 1950s was understood in terms of participation in industrial labour, even in construction and heavy industry. As well as the obese, skinny figures also signalled manual inefficiency and contradicted the dominant values.

This relationship of the bodily form to an ideal remained the same approximately until the mid-1950s, though its remnants persisted even after that. At the end of Socialism's golden years, in 1955, the sociologist and health nutrition journalist Božena Solnařová published an article in the magazine *Výživa lidu* called 'The Diet of Young Girls and their Determining Influences'. In the article she not only outlined the above-mentioned ideological notions, but also pointed to some initial complications in enforcing such ideas among the general public. The author sharply criticized the spread of attempts to follow different diets for aesthetic reasons, which she perceived as a dangerous ideological import from the West. She expressly talks about the 'fabricated dictate of egoistic fashion designers, which is imprinted in the current aesthetic ideal for men and women through some kind of mass psychology'. Although the article can be considered a type of warning against imitating foreign models, the situation at home did not appear to be so threatening at the time. Solnařová stated that

> it is not at all possible to compare our own girls, not afraid of hard labour and physical strain and enjoying all gifts of life, with these poor creatures, bereft of one of the justified joys of every human being (i.e. girls from capitalistic countries who drastically limit the intake of fats in their diet). Our girls set out on a successful path to be the ideal woman, combining charm, health and strength...[5]

---

4   In the 1960s the depiction of members of the bourgeoisie as fat people was criticized as materially incorrect. At a congress of the Society for Rational Nutrition in 1967, for example, J. Cvekl said: '...when a capitalist was painted, he was painted with a huge belly and thin legs [notice the mentioned of thin legs – a symbol of doing nothing – author's note]. However, whoever has seen today's capitalists knows that they do not look like that. Most of today's tycoons of the capitalist industry eat very rationally. Most of them are on strict diets and are mostly very slim.' Quoted in Adamec 1968: 46.

5   Solnařová 1955: 44. As an example of unhealthy tendencies arising from the 'unconditional docility of the fashion line', the author refers to the prewar and postwar figures of an unnamed 'French fashion magazine of worldwide publicity'. According to her, the efforts to 'find its place in the erotic life, to draw attention to itself and to attract the other sex' are much stronger in a society where 'an attractive appearance is one of the few opportunities of social promotion and successful life for a woman,' i.e. in capitalist society.

Interwar protagonists of temperance in eating, often associated with vegetarianism, were similarly criticized sharply, as were drastic Western slimming diets. The religious tradition of fasting was also attacked in this respect.[6] However, eating in Czechoslovakia started changing visibly around the mid-1950s. In 1953 the rationing system ended, though it was accompanied by stringent monetary reform and huge problems with the food supply and returned the following year. Nevertheless, the average energy value of food consumption per person in the Czech lands started growing continuously and in 1956 exceeded 3,000 kilocalories per day.[7] The question of body weight again became the centre of attention.[8] This gave rise to so much concern over excess bodyweight that it became a popular image of the Czechs much used by journalists. When Božena Solnařová was writing her article, the Congress of the Dietary Section of the Society for Rational Nutrition[9] classified obesity as one of the most common diseases that threatened public health.[10] The statistics from the 1960s hinted that the problem involved

In this respect Solnařová quotes an observation by one of the leading official ideologists ideologists of the day, Zdeněk Nejedlý, President of the Czech Academy of Sciences. The author admits, however, that even in the domestic environment, on the basis of a survey conducted in 1948, fashionable clothing was a big competitor to food for girls at boarding schools – the girls often used lunch money to buy fashionable clothes.

6   See Šíma 1950. Šíma talked derogatively of 'popes and half-popes' who 'preached the merits of cereal half-hungry nutrition', who 'indirectly served the bandit classes and their agrarian, pricing and nutrition policy'. He perceived the low consumption of animal proteins arising out of the mentioned diets as one of the 'tools of universal exploitation of the working classes'. Religious fasting was evaluated in a similar negative way. At that time Šíma was an exception in the professional discourse on nutrition with his ideological attacks, even though Mašek also ironically noted in a report from the Second International Congress of Dietetics, held in Rome in 1956, that 'many people are puritanical about fatness and perceive it as a big sin – non-temperance'. See Mašek 1956: 146. It should be added that that in his other studies Mašek criticized Protestantism for its negative approach to food.

7   This is, however, gross consumption, i.e. including food losses during processing. As estimated, the net energy intake did not reach 3,000 kilocalories per person per day in mid-1960s. During the early 1950s the energy intake per person/day decreased slightly compared with 1950, but started increasing in 1953. See *Statistická Ročenka ČSSR 1961*, Prague, 1961: 392 and Franc 2003: 235–6.

8   In her work Marta Vamberská refers to a research by V. Kapalín that showed a direct and close correlation between the number of obese children and the current economic situation (ending of rationing and lower prices of food). See Vamberská 1963: 7. For comparison with Germany see Thoms, in the present volume.

9   For the Society of Rational Nutrition see Franc, M. 'National tradition or happy tomorrows? The dilemma for Czech nutritional science in the 1950s and 1960s' in ICREFH VIII, 269–76.

10   Final resolution of the Fourth Congress of the Dietary Section of the Society for Dietary Nutrition on 10 and 11 October in Luhačovice 1955, 154.

one-quarter or one-third of adult men and more than one-half of all adult women.[11] The scope of research and obesity-related promotional campaigns grew rapidly.[12] The year 1960 can be described as the peak of this trend, because that year saw a national congress on the issue of obesity and excessive weight.[13] In the same year the annual promotional campaign Nutrition and Health was held with the pointed slogan 'With Temperance from a Rich Table' and with the even more telling sub-motto that accompanied anti-obesity campaigns for several decades to come: 'Getting Fat Means Getting Old'. In the following years, the overall energy value of the average diet decreased, but this was due to economic problems and failures in supply, especially of animal products, rather than successful medical propaganda.

In the course of the 1950s, the ideological discourse that shaped discussions on obesity understandably started to alter. First and foremost, there were efforts to interpret the growing number of obese people as evidence of the rising standard of living in the regime established after 1948. This was associated not just with an abundance of cheap food,[14] but also with the reduction of heavy manual labour.[15] This approach was also related to the assumption that the issue of slimming and keeping to a diet did not have anything to do with workers in capitalism. Although Božena Solnařová shed tears over the poor creatures in Western countries who had to keep to drastic diets to maintain their jobs, in 1957 the main newspaper of the Communist Party, *Rudé Právo*, which aspired to become the regime's official mouthpiece, aggressively attacked the British Labour newspaper, the *Daily Herald*. From the 8 March 1957, a series of articles on a new (and of course fantastically efficient) slimming diet was announced on the *Daily Herald*'s title page. *Rudé Právo* linked this announcement with reports on price increases involving milk and school lunches and stated resolutely that 'workers in England do not suffer from a problem of getting fat. Rather they are worried about what they can give their children to eat.'[16]

---

11   Mašek 1967: 16. Some authors considered Mašek's definition of obesity very strict, but his conclusions were more or less supported by other surveys. See Hejda and Ošancová 1969: 19–20.

12   Based on a public opinion survey from 1958, over 50 percent of all people considered 'not eating excessively' a significant requirement for healthy nutrition. It was the most common general requirement for health when eating. See Adamec 1961: 1124.

13   Hrušková 1961.

14   Novák 1960: 153. In 1967, 36 per cent of people mentioned a high standard of living as a reason why people in the Czech Republic ate a lot; 9 per cent mentioned 'enough food' in this respect. Adamec 1968: 38–9.

15   On the other hand, there was faith that socialism would find a way to prevent a decline in physical strength and endurance. Mašek 1960: 162. The socialist society should have been the only society to enable a change in eating habits by means of a more suitable assortment of foodstuffs. Baláž 1961: 21–2.

16   Jh, 1957, 3.

Next to threats of hunger faced by West European workers,[17] popular topics at that time included insufficient and unhealthy nutrition in many third-world countries, usually accompanied by convictions that, after liberation from colonial rule, embracing socialism would solve all problems quickly.[18] No wonder that in light of these reports, the epidemic of obesity seemed to usher in improving conditions for the general public.[19] Under these circumstances gaining weight seemed to be something really luxurious, even though sinful.[20] We can assume that these convictions were shared at least by some politicians who adhered to opinions typical of the workers' environment in the interwar period as regards issues of lifestyle.[21] Moreover, a large part of society still maintained memories of the war and post-war insufficiency, which obviously increased the importance of food in the social hierarchy of values.[22] Expenditure on foodstuffs in the total household budget was very high compared to West European countries or the USA.[23] As late as in 1967, during a public-opinion poll focused on the ranking of desirable lifestyle values, the requirement to eat well was placed second, after the general wish to 'live well.'[24] As medical opinion increasingly dominated the anti-obesity

---

17   See also Kovářová 1956: 158. However, similar information quickly disappeared from expert statistics in the second half of the 1950s. Franc 2003: 202–10.

18   Doberský 1965: 69–76. It is a secondary-school textbook for dietary nurses. Among other things the text says that, in accordance with the Statistical Yearbook, 10 per cent of the population of Great Britain lived in conditions of undernourishment. Another interesting statement is that 'more than a third of the world has already overthrown capitalism and, led by the Soviet Union, is heading rapidly towards a socialist and Communist society. The colonial dominion in Asia and Africa has crumbled and will soon disappear completely. The oppressed peoples of Central America are also being liberated. The world of hunger is disappearing.' Characteristically, the text is accompanied by pictures of undernourished children from Vietnam taken in 1960.

19   In a public opinion poll conducted in 1967, about one-quarter of the respondents associated a high standard of living with an obese or corpulent figure. Adamec 1968: 47.

20   Josef Charvát, Preface to Doberský, Doleček and Šonka 1967: 9.

21   See the speech of J. Cvekl at a Congress of the Society for Rational Nutrition, which included the statements of a Communist official from the 1950s: 'Comrade, but that's a sign of socialist well-being, we are fighting for a higher standard of living. And when people are – as we say – plumper, it is a sign of this standard of living.' Quoted by Adamec 1968: 46.

22   The high consumption of food was justified by some respondents as a residual echo from the war as late as in a public-opinion poll in 1967. Adamec 1968: 39.

23   About the mid-1960s a worker's family allegedly spent over 50 per cent of its income on food. This figure is given by Doleček in Doberský, Doleček and Šonka 1967: 248. Adamec 1968: 40 also refers to the data from the Central Popular Inspection and Statistics from 1966, mentioning the words of the then Deputy Prime Minister and father of the economic reform, Ota Šik, uttered in a TV debate. However, the official yearbook contains a different figure (44.1 per cent). The causes of the dispute are unclear and are probably related to the different methodologies for calculating the given indicator.

24   Adamec 1968: 66–7.

campaign, doctors fought against these notions often using ideological arguments. They especially criticized the excessive emphasis on food as an expression of consumerism, which led not just to the neglect of sports and cultural activities, but also limited public activities. This seemed to be a very serious offence from the regime's point of view, which described the emphasis on nutrition as a residue of the past or as barbarism.[25]

In the more relaxed atmosphere of the 1960s, the notion of a direct relationship between the growth in the number of overweight people and the rising standard of living was challenged by some experts who claimed that, in the conditions of a socialist country, obesity remained a disease of the poor, because its inception was caused by an unsuitable choice of food influenced by a lack of money. This meant, in the Czech environment, that poor people preferred food rich in sugars, especially various types of bakery products. It was clearly formulated by Josef Mašek, perhaps the most important Czech expert in nutrition, who wrote in his book *Člověk, společnost a výživa* [Individual, Society and Nutrition], published in 1971, that ageing women especially

> ...are forced into the arms of obesity and the related shortage of vitamins, because they are compelled to eat cheap foodstuffs involving baked products and poorly coloured, but very sweet coffee.[26]

In general, the price structure, especially the fact that foodstuffs rich in sugars were much cheaper than protein foods, was criticized heavily in the 1960s.[27] From this point of view those campaigning against obesity resisted the excessive deregulation of the price system, especially in the second half of the 1960s, and sought regulation justified by health considerations. In doing so they emphasized that a society directed towards communism should cope with all so-called exogamous effects such as the increase in obesity. This was not just a question of preventive monitoring and control of retail prices, but other areas, too. Much criticism was made of the contemporary network of sport facilities and clubs,

---

25　Šonka and Přibylová-Čárková 1968: 9. Many works on the issue of obesity deal with the issue from a historical point of view in a short retrospective. Very frequent are mentions of corpulent old-age Venus statues. At the same time some authors typically emphasize that 'for ancient cultural nations obesity was definitely not a sign of beauty and harmony'. Vamberská 1963: 5. See also Adamec 1968: 42, who cites a speech by V. Křížek. Novák says, on the other hand, that 'excessive eating and obesity were requirements among primitives, whether they lived in caves or in imperial palaces.' Novák 1960. There were signs of a historical approach in Soviet literature, too – see Yegorov and Levitskiy 1964: 13–18.

26　Mašek 1971: 158. The social situation of older women was often very dramatic in Czechoslovakia. Since many of them had not worked for most of their lives, they were entitled to very low, so-called social, pensions.

27　Doleček in Doberský, Doleček and Šonka 1967: 247–8.

where excessive attention was allegedly paid to the training of top athletes and also to the public catering facilities.[28]

As part of the fight against obesity, the Czech national cuisine was frequently criticized. This, however, was controversial, as the Czech cuisine was a part of the officially supported popular traditions. Most experts resolved this disparity by referring to different living conditions existing when the national cuisine developed, in particular, the prevailing heavy demands of physical activity.[29] The urgent criticism of shortcomings in contemporary industry or the distribution network may have seemed equally dangerous. It was usually accepted, but its practical implications were minimal.[30] The economic reformers of the 1960s especially emphasized that high consumption of food was often the result of long-term deficiencies in the supply of other types of goods.[31]

In the second half of the 1950s physicians tried to take over the whole discussion on the issues of nutrition. In particular, a rather small and narrowly defined group of specialists focused on obesity as a cause or risk factor in many different diseases, for example cardiac conditions and diabetes mellitus.[32] Conversely, slimming diets for aesthetic reasons were still considered rather doubtful, especially when many of these fashionable adjustments of eating habits revealed inspiration in Western consumerism as, for example, the formerly very popular 'Hollywood diet'.[33] The most important figure in this opposition to any such fashion was probably Božena Solnařová who, in 1967, at a time of full liberalization, described with clearly stated irony, a diet adopted from the French women's magazine, *Marie-Claire*.[34] However, even Solnařová admitted at that time that

---

28   Vamberská 1963: 92–5.

29   On the issue of relationship to the national cuisine, see Franc 2003.

30   Franc 2003: 170-1. The hopes of some nutrition experts given to the development of catering in companies and schools were also unrealized. Compared with their original expectations, the structure of the meals often differed from the requirements of rational nutrition much more than did home-made food. Doberský, Doleček and Šonka 1967: 247.

31   Adamec 1968: 40, where Ota Šik, the former Deputy Prime Minister and author of the Czechoslovak economic reform, is quoted.

32   They included, in addition to the Director of the People's Nutrition Research Institute, Josef Mašek, also Jiří Šonka, Mašek's subordinate, Přemysl Doberský, and physician Rajko Doleček, who later, in the 1970s, became the most famous campaigner against obesity. An exception in this company was the long-term employee of the Central Health Education Institute and sociologist, Čestmír Adamec, who backed his reasons with the opinions of medical professionals. The already mentioned specialist Josef Charvát, author of a famous diet from the interwar period, commented on the issues very little at this time.

33   For differences between rational nutrition aimed at higher working efficiency and a diet focused on slimming, see Wildt 1996: 206–8.

34   Solnařová 1967: 40. This refers specifically to an interview with Ms Dorian Leigh, the owner of a school and agency for fashion and advertising models.

our current times, with their ideal of speed, sports performance, sense of simple lines, with its cult of youth and slimness, are obviously heading, at least in theory, to light, simple nutrition, and understand the need for similar advice from medical professionals.[35]

Instead of importing slimming plans from the West, Czechoslovak physicians especially recommended a diet developed by the People's Nutrition Research Institute, which followed the interwar dietary system of Josef Charvát. For bolder experimenters, they offered a diet of so-called contrasting days by Přemysl Doberský. On these contrasting days slimmers were allowed to eat virtually only one type of food – e.g. whipped cream or apples. Interestingly, however, in the 1960s there was much scepticism about a diet by the Soviet doctor, Yegorov, which according to Czechoslovak medical authorities did not maintain the overall nitrogen balance normally considered a key parameter for evaluating the quality of individual slimming diets.[36]

At the same time, medical authorities were compelled to respond to Western diets acquired from various women's magazines. This especially applied to the strict 'Hollywood diet' and the similar diet from the Mayo Clinic which were both very popular in the Czech environment. Compared with them, however, the regimens prepared by domestic physicians had one undisputed advantage – they mostly took into account the food offered in Czech shops. For instance, in the 1960s, both American dietary regimens mentioned required a large amount of citrus fruits, the supply of which was very limited in Czechoslovakia. In his evaluation, Přemysl Doberský restricted himself to saying that both diets presented 'significant limitation' for obese people who could only follow them 'probably with great difficulty'. There was no clearer statement, though, most probably because local medical professionals did not test the effectiveness of either diet clinically.[37] The

---

35 Solnařová and Hrubý 1967: 213. Note the mention about sport performance – athletes required sufficient nutrition. The admiration for manual labour was replaced by emphasis on physical activity in the form of sporting achievements.

36 Doberský, Doleček and Šonka 1967: 104–5.

37 Ibid, 105. See also Doberský, Horáčková and Šejbalová 1970: 234, where one-sidedness and difficult compliance with a diet are discussed. On the other hand, it is accompanied by a painting of a very attractive and 'Western-looking' slim woman. Much sharper criticism was given to the so-called 'points' diet in the next edition of the same work; the diet was labelled as unscientific and was not even described. Doberský and Horáčková 1976: 172. There were similarly sharp attacks against the 'points' diet by Josef Mašek in 1971. This author was especially amazed that a lot of points were given, for example, to tomatoes and bananas, while whisky and cognac could be drunk without any limitation. Nobody paid any attention to the Hollywood diet or the Mayo Clinic diet. See Mašek 1971: 157–8. The main criticism in the dietary literature of the 1960s was focused not on any innovation, but on the so-called Schrott diet, originating from the first half of the nineteenth century. The Schrott diet was used for a long time in the leading Czech healthcare facility for the obese in Dolní Lípová and consisted in dehydrating the body and advocating the use

growth of fashionable diets showed that the medical propaganda favouring the fight against obesity as a struggle for higher efficiency and prevention of different lifestyle diseases was not as successful as expected, in spite of all its official support. However, its results looked imposing at a glance. This can be best shown by an extensive public-opinion poll conducted at the end of the period in question, in 1967. The survey was conducted by the Institute of Public Opinion Research, which deliberately chose a topic that did not stir up strong political emotions. According to the survey, only 10 per cent of all respondents mentioned aesthetic problems as a disadvantage of obesity, while propensity to diseases was mentioned almost three times as often. In addition, when asked directly if the respondents thought obesity was the cause of various diseases, more than three-quarters of them answered 'yes'.[38] On the other hand, three years later the well-known actress and author of many texts and programmes on food, Jiřina Šejbalová, wrote in the preface to a book on slimming that:

> putting aside the small percentage of people who have accepted the view of modern medical science that obesity is actually a disease, the reasons for fighting excessive weight are in short as follows: (a) existential [the author refers to professional reasons, e.g. for actresses, models. etc.], (b) vanity, (c) regards for one's own wardrobe.[39]

Should we refute the relevance of this opinion forthwith and without any further considerations, giving regard to the specific environment from which the author came? I think this would be an inadmissible simplification – similar opinions can be found voiced by other authors, including persons from the medical environment.[40] In my opinion the results of the public-opinion poll mentioned above could also be questioned. The validity of medical propaganda was very strongly entrenched in people's minds: it was seen as the official view of the issue which understandably showed up in opinion polls. For that reason it should be borne in mind that the answers of the respondents often reflected the public approach to the issue rather than their actual opinions. It was probably the first large field survey of public opinion ever conducted in Czechoslovakia.

---

of wine and dancing. After prolonged disputes, the Schrott diet was abandoned in Dolní Lípové in the mid-1960s. Doberský, Doleček and Šonka 1967: especially 106–7, 151–2 and 266–7, Štolbová 2000: 40.

38 Adamec 1968: 21, 23.

39 Doberský, Horáčková and Šejbalová 1970: 14.

40 See Matek and Sobotková 1963: 11, where we can similarly read that 'obesity is most often judged from an aesthetic point of view, which is especially dominant for women, because they anxiously observe their weight and try to remove any unwelcome fat, to make sure they are as close to the fashionable ideal of beauty as possible, even if it is very different from the normal bodily weight.' See also Baláž 1961: 18.

The fact that aesthetic considerations played a more important role in the fight against obesity than leading medical professionals would be willing to admit, is supported by the development of one interesting institution that dealt with, among other things, obesity treatment.[41] This was the Institute of Cosmetics, established in the late 1950s in Prague, which later extended its services to cover many other cities. Originally it was actually designed as a purely medical institution but services related to the care of the woman's body, such as cosmetics and hairdressing soon prevailed among its activities. A commercial consultancy for overweight people was quickly set up in the Institute of Cosmetics. Even though the consultant there was Jiří Šonka, one of the most famous experts of those times and the author of many specialized publications, an overwhelming majority of the clients did not even try to conceal that they were coming especially for aesthetic reasons. Under the veneer of the officially sanctioned medical fight against obesity (which stressed absence from work due to illness amongst obese workers and the cost of their disability pensions), a cult of the slim body was slowly evolving, influenced heavily by Western culture. In the course of time many doctors started to use aesthetic factors for their promotion of rational nutrition. No doubt this was also due to the ideological change in approach to the consumer society which Radoslav Selucký, the popular Czech economist of the late 1960s, considered desirable and fully compatible with socialist society.[42]

According to contemporaries, the famous model Twiggy was very popular even in the Czech environment of the 1960s; however, she did not have any Czech counterparts.[43] As opposed to West European standards, domestic 'celebrities' did not take much part in the discussions on obesity. Some of them hinted, almost conspiratorially, that they did not care for slimness too much and cast doubt on the need to follow medical recommendations strictly. This approach was based to a certain degree on the specific position of medical propaganda, which constituted a part of the official, governing discourse – in practice, one that could be rebelled against without any significant problems, much as popular journalists portrayed the Czechoslovak people as overweight, despite government policy

---

41   Especially in the second half of the 1960s some doctors admitted the importance of aesthetic factors. See Rath, 1967, 70–1, where we can find the sentence 'Not only expressly medical, but also social and aesthetic reasons speak clearly against obese people.' Similarly, Rajko Doleček mentioned that doctors should also convince obese patients of the necessity of reducing their weight on the basis of social and aesthetic aspects. Doberský, Doleček and Šonka 1967: 264.

42   Selucký 1966: especially 65.

43   In the public-opinion poll of 1967, 4.5 per cent of the male respondents chose a very slim figure of a woman as aesthetically the most attractive on a scale from a very obese figure to a very slim one. A slim figure was much more popular (65.7 per cent), just like a 'normal' figure (28 per cent). As for women, they preferred men of a normal build to slim (56 to 40.3 per cent). Interestingly, an obese man was preferred by more women than a very slim figure (1.9 to 1.5 per cent). Adamec 1968: 108. These data applied to Czech lands; in Slovakia men and women liked fatter partners more often.

to the contrary. In a sense these were easily achievable 'forbidden fruits'. Jiřina Šejbalová (1905–81), the Czech actress, was such a case. Her notes, often not too respectful, paradoxically accompanied the popular book *Abychom netloustl* [To Prevent Obesity].[44] This trend continued in the 1970s and partly also in 1980s.

The Soviet intervention and the subsequent 'normalization' period did not represent a radical turnaround in the fight against obesity. It continued, though less emphasis was given to obesity as a disease of poverty. The long-standing validity of the reasoning used in the 1970s and 1980s produced little medical success. Although they had tremendous scope for propaganda, they only had a minimal impact on practical decisions, just as in the 1950s and 1960s. In addition, they could not count on support from the inspiration of Western consumerism that they had received in the 1960s. Once again, Western consumerism was approached with a high degree of suspicion and its shadow fell upon any initiative and everyone associated with it.

References

Adamec, Č. 'Poznatky z průzkumu zdravotního uvědomění obyvatelstva v otázkách správné výživy' [Survey of Public Awareness of the Questions of Healthy Diet], *Časopis lékařů českých* 36, 1961, 1121–7.

Adamec, Č. et al. *Veřejné mínění a otylost* [Public Opinion and Obesity], Prague, 1968.

Baláž, V. *Tučnota* [Obesity], Bratislava, 1961.

Doberský, P. *Nauka o výživě a dietetice* [Science of Nutrition and Dietetics], Prague, 1965.

Doberský, P., Doleček, R. and Šonka, J. *Léčení otylosti* [Curing Obesity], Prague, 1967.

Doberský, P., Horáčková, J. and Šejbalová, J. *Abychom netloustli* [To Prevent Obesity], Prague, 1970.

Doberský, P. and Horáčková, J. *Abychom netloustli* [To Prevent Obesity], 2nd ed. Prague, 1976.

Franc, M. *Řasy, nebo knedlíky? Postoje odborníků na výživu k inovacím a tradicím v české stravě 50. a 60. letech 20. století* [Algae, or Dumplings? Attitudes of nutrition experts towards traditions and innovations in nutrition in the Czech lands in the 1950s and 1960s], Prague, 2003.

Hejda, S. and Ošancová, K. 'Odraz spotřeby potravin na zdravotním stavu' [Food Consumption and Health], *Výživa lidu* 24, 2, 1969, 18–21.

---

44   Cf. Doberský, Horáčková and Šejbalová 1970: 13–15. What is interesting is that Jiřina Šejbalová was not mentioned in the next edition of the same work. However, the validity of medical opinions was generally confirmed, while it was hinted that they would not have to follow them so strictly.

Hrušková, J. 'Celostátní sjezd o otylosti – Karlovy Vary, 27 – 29 Sep 1960' [All-State Congress on Obesity], *Časopis lékařů českých* 27–8, 10, 1961, 885–8.

jh, 'Není nad správnou "kontrolu diety"' ['Control of Diet' – Truly a Crucial Issue], *Rudé Právo* 14 Mar 1957, 3.

Kovářová, A. 'Naše účast na zdravotnickoosvětové konferenci v Římě' [Our Participation at the Health Awareness Conference in Rome], *Výživa Lidu* 11, 11, 1956, 158–9.

Mašek, J. 'II. mezinárodní kongres dietetiky v Římě' (11. – 14. září 1956) [2nd International Congress of Dietetics in Rome (11–14 September 1956)], *Výživa Lidu* 15, 10, 1956, 146–7.

Mašek, J. 'Výživa a růst mládeže z hlediska výživy obyvatelstva' [Nutrition and Growth of Youth from the Perspective of Nutrition of Population], *Výživa Lidu* 10, 11, 1960, 161–3.

Mašek, J. 'Zdravotnický význam otylosti' [Medical Importance of Obesity], in Doberský, P., Doleček, R. and Šonka, J. (eds), *Léčení otylosti*, Prague, 1967, 15–17.

Mašek, J. et al. *Člověk, společnost a výživa* [Individual, Society and Nutrition], Prague, 1971.

Matek, J. and Sobotková, M. *Udržte si štíhlou linii* [Stay Slim], Prague, 1963.

Novák, M. 'Otylost' [Obesity], *Výživa Lidu* 15, 10, 1960, 152–4.

Pevzner, M.I. *Základy léčebné výživy* [Basis of Medical Nutrition], Prague, 1952.

Rath, R. 'Otylost a její následky' [Obesity and Its Consequences], *Výživa Lidu* 22, 5, 1967, 70–71.

Rogl, V. *'Po únoru 1948 se zpívalo: 'Máme Martu prdelatou...'* [After 1948 They Sang 'We Have Fat Marta'], Naše noviny 1/2005. Available at: http://www.nasenoviny.net/detailn.php?nn=1/2005&id_cl=1_2005_23&roknn=noviny05 [accessed: 20 April 2009].

Selucký, R. *Člověk a jeho volný čas. Pokus o ekonomickou formulaci problému* [Individual and Leisure, a Economic Perspective], Prague, 1966.

Solnařová, B. 'Výživa mladých dívek a vlivy, které ji určují' [The Diet of Young Girls and their Determining Influences], *Výživa Lidu* 10, 3, 1955, 43–5.

Solnařová, B. 'Ani kritika, ani rada' [Neither Critique, Nor Advice], *Výživa Lidu* 22, 3, 1967, 40.

Solnařová, B. and Hrubý, J. *Výživa jako ekonomický problém* [Diet as an Economic Problem], Prague, 1967.

Statistická ročenka ČSSR 1961 [Czechoslovakia's Statistical Yearbook 1961], Prague, 1961.

Šíma, J. 'Spotřební zvyky ve výživě z hlediska sociálně ekonomického' [Consumption Habits in Nutrition in Social and Economic Perspective], *Výživa Lidu*, 5, 4, 1950, 73–5.

Šonka, J. and Přibylová-Čárková, M. *Dieta při otylosti* [Diet against Obesity], 5th ed. Prague, 1968.

Štolbová, E. *Život s nadváhou* [Life with Obesity], Prague, 2000.

Vamberská, M. *Léčení otylosti u dětí a mladistvých* [Curing Obesity of Children and Youth], Prague, 1963.

Wildt, M. *Vom kleinen Wohlstand. Eine Konsumgeschichte der fünfziger Jahre*, Frankfurt am Main, 1996.

Yegorov, M.N. and Levitskij, L.M. *Ožirenije* [Obesity], Moscow, 1964.

Závěrečné usnesení IV. sjezdu dietní sekce SRV ve dnech 10. a 11. X. v Luhačovicích [Final Resolution of the 4th Conference of the Dietary Section of the Society for Rational Nutrition on 10 and 11 October in Luhačovice], *Výživa Lidu* 10, 11, 1955, 154.

Chapter 15

Separated, But Sharing a Health Problem: Obesity in East and West Germany, 1945–1989

Ulrike Thoms

The Beginning of Nutrition Policy after 1945

When Germany signed the unconditional surrender in May 1945, the occupiers were shocked by the signs of undernutrition they observed in the people. Food policy at that time meant rationing and rations remained low until 1949. Undernutrition was the main problem until rationing ended in 1951 in the Western zone and in 1958 in the East (Table 15.1).[1]

Table 15.1   Comparison of the Energy Value of the Lowest Official Rations in the Eastern and Western Occupation Zones of Germany 1945–1949

Date (Month/Year)	British and American Zones (kcal/day)	Soviet Zone (kcal/day)
5/1945–9/1945	1,000–1,200	–
9/1945–2/1946	1,380–1,580	940–1,114
3/1946–10/1946	1,015–1,235	940–1,114
10/1946–3/1947	1,330–1,550	1,114–1,288
3/1947–3/1948	1,100–1,400	1,338–1,520
4/1948–6/1948	1,350–1,760	1,338–1,520
7/1948–9/1948	1,830–1,990	1,338–1,520
10/1948–3/1949	1,830–1,990	1,528–1,588
4/1949–5/1949	1,830–1,990	1,528–1,588

Source: Boldorf 1998, 80.

---

1   There is a large body of literature on the rationing question. For a comparative perspective see Gries 1991.

From the 1950s onwards, the situation changed. Food consumption and energy intakes rose remarkably and notes on the role of obesity in the aetiology of diabetes and arteriosclerosis began to be published.[2] Medical staff published graphs and tables like Figure 15.1, which showed how the body weights of patients admitted to a Berlin clinic increased between 1946 and 1950 by different age groups. Above all, these figures demonstrate how people regained weight losses in the aftermath of war. Nevertheless, as early as 1952, alarming articles on the increase of obesity were published by medical staff specializing in the treatment of diabetics. They were strongly orientated towards the USA, where similar developments had already led to a campaign against obesity. It is important to note the role of Americanization, that is, the strong influence of American culture and science on Western Germany after 1945. In the case of obesity this process was driven by the investigations of American life insurance companies, namely the Metropolitan Life Insurance tables. Although criticized in detail, the overall argument was that overweight people's life expectancy was lower and their risk of cardiovascular disease was higher. The graphs are taken from a similar study in which people were weighed to assess how many of them had a weight more than 10 per cent above the Broca Ideal Body Weight. The graphs obtained in this way clearly showed an increase in body weight, which correlated with increasing energy intakes.[3]

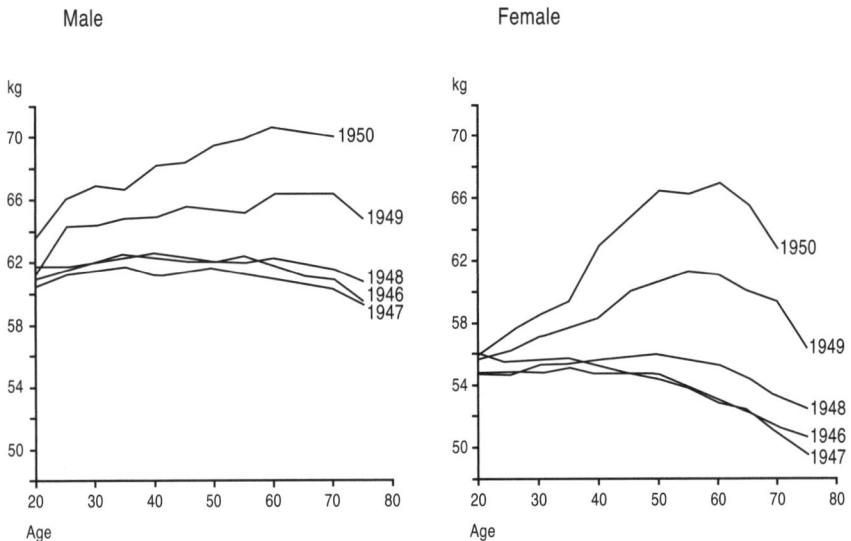

Figure 15.1   The Weight of Adults in Berlin According to Age Groups 1946–1950

Source: Redrawn after Meier 1956.

---

2   See for example: Bertram 1950; Damm 1951.
3   Grosse-Brockhoff 1953; Berger 1973.

Although it is worthwhile discussing the different indices and their history as an important tool of the biopolitics of today, this opens up more questions than it answers: for example, how appropriate these indices are to measure the potential risk of food-related illness.[4] Instead, this chapter will concentrate on the diabetes question. The relationship between being overweight and diabetes has been known for a long time. It may be questionable how far obesity is a risk leading to the incidence of type II diabetes but, when diagnosed, the condition improved when body weight was reduced. As a severe disease, it will – if left untreated – lead to death or premature death. This correlation seems to be quite clear; however the number of newly observed cardiovascular diseases and deaths following strokes has been going down over the years despite the increase in average weights.[5] So diabetes may be a more reliable indicator. In fact, the prevention of diabetes has become a main target of public-health policy and nutrition education, since diabetes causes high costs within the system of social insurance.[6]

The following discussion will examine how both German states, with their diverging political and public health systems, reacted towards the alarming increase in the number of overweight people and diabetics, and how they organized nutritional information and health education in order to prevent the further development of what is considered a major public health problem of today.

Obtaining Reliable Data

Statistics play an important role in the emergence of health policy. Population statistics and epidemiological data are important tools for governments as they help to identify problems and their extent, to classify the groups of people involved and to develop remedial policies.[7] Weighing random samples of people on a systematic basis is a recent development, but assessing the incidence of diabetics goes back to World War I. Later, as diabetics had to have a special permit from a physician in order to get extra rations when rationing began in 1939, medical specialists calculated the number of cases from the numbers of permits issued.[8] Using this method, a first nationwide survey was undertaken in 1941.[9] It showed a decrease in the number of diabetics from 2.2 to 1.5 deaths per thousand from the

---

4 Among the large number of publications on this topic see Schmidt-Semisch 2008, Gilman 2008; Thoms 2000; Schwartz 1986; Beller 1977; and Csergo, Chapter 12 in this volume.

5 Hauner, Köster and Schubert 2004.

6 WHO 2000, reported a worldwide obesity epidemic and started a global campaign against it.

7 Classical works in this field are Hacking 1990; Porter 1988.

8 See for example Umber 1940.

9 Bundesarchiv Koblenz, NL 329/17. Due to World War II, the study was published only in 1960.

beginning of rationing in 1939.[10] This confirmed the findings from World War I that a restricted diet resulted in a reduction in the incidence of diabetes. It meant a paradigm change in public health concerns about nutrition, which had been primarily orientated to the minimum of physical needs. This view had dominated nutrition policy and education throughout the nineteenth century and during the times of scarcity during and after World War I and the interwar economic crisis. The weight tables and indices which had been used to find out those whose health was endangered by food shortage could be used to analyse obesity from the 1950s onwards.[11]

Two German States – Two Models of Health Policy

It is important to note that scientific co-operation between the Eastern and Western parts of Germany worked more or less well until the Berlin Wall was built in 1961, so that each part of Germany participated in the scientific findings of the other. Nevertheless there were important differences in regard to health policy.

Health Policy Eastern Style

Health policy was part of the German Democratic Republic's constitution from 1949.[12] This commitment to health resulted from the ideal of a socialist utopia and was part of the concept of social hygiene which stressed the state's responsibility for the population's health.[13] Health was seen to be the consequence of life circumstances and it was the task of politicians to ensure healthy living conditions.[14] The active Socialist was responsible for his health, as the wellbeing of society depended on his work.

In East Germany surveys of the number of diabetics were undertaken as early as 1950. Their main task was to identify diabetics in order to begin treatment as soon as possible and to limit the need for insulin. Insulin was not produced in the GDR but had to be imported and thus used up scarce foreign currency. Driven by these economic considerations, the GDR developed a highly centralized system of care for diabetics in which prevention and nutrition education were important. At the State level, a Diabetes Committee was founded in 1956. It had sub-committees which organized surveys, observed the overall development of diabetes and the measures taken. It worked out guidelines for diets and policy at

---

10   Bundesarchiv, Koblenz NL 329/17, 72.

11   Thoms 2004b.

12   See Article 35 of the Verfassung der Deutschen Demokratischen Republik, 1981, 32 and Elkeles et al. 1991.

13   Elkeles 1991; Stöckl 2000; Schagen and Schleiermacher 2006.

14   Niehoff and Schrader 1991: 51–4. In fact politics violated these rights in many respects, e.g. in the case of environment policy.

the national level, as well as central therapeutic guidelines which were published in 1973 and 1974. Dispensaries cared for the diabetics and collected statistical and very detailed medical information. As the diabetics were expected to go to the dispensary every four to twelve weeks in order to get their insulin or pills and to be instructed by special diabetes advisers, they could be surveyed, followed up, and kept under constant scrutiny easily. This system was made more efficient by combining the regular tuberculosis screenings of the whole populace with diabetes tests in order to detect diabetics as early as possible. In 1958 the Central Diabetes Register (*Zentrales Diabetesregister*) was introduced, which listed all known cases of diabetes, including newly detected cases as well as deaths from diabetes. Only the more complicated cases were sent to the special clinics.[15]

The Institute for Nutrition in Potsdam-Rehbrücke played an important role. Founded in 1948 it was conceptualized to be the GDR's central and leading research institute in nutrition and dietary affairs.[16] When it became an Institute of the German Academy of Sciences in 1957, its constitution stated that it took over 'the task of introducing scientific ways of nutrition, of conducting measures of education, consultation and steering as well as organizing training seminars, courses and lectures'.[17] Consequently, it held an important position in the centralized economic and health policy of the GDR. From 1958 – which means from the very end of rationing – there were warnings about the severe consequences of overeating, not only by the Institute for Nutrition, but even by the leading members of the Communist Party.[18] In the context of the declared aims of a unified social and economic policy, economic, organizational and educational measures were integrated into an overall concept, which was based upon the firm belief in the possibilities of science as an indispensable part of the socialist society.[19]

One way to follow such a holistic strategy was to subsidize communal feeding (*Gemeinschaftsverpflegung*), that is mass catering, which fulfilled several functions: females and mothers, who were expected to work, were partly relieved from their household duties,[20] and people were provided with all the nutrients and vitamins necessary to balance the deficiencies of an insufficient morning or evening meal.[21] Finally, it would have such educational effects as the 'Outline of Health Education' stated in 1973:

---

15 Bruns et al. 2004, Dittrich 1986.

16 For the early history of this institution see: Linow, Lewerenz and Möhr 1996; Thoms 2009.

17 Ordnung der Aufgaben und der Arbeitsweise des Instituts für Ernährung der Deutschen Akademie der Wissenschaften zu Berlin vom 18. September 1957, cited in Ketz 1996: 210.

18 Ulbricht 1963: 28.

19 Semmler and Ketz 1973: 16.

20 Scheunert 1955: 104.

21 Scheunert 1955.

Collective feeding in companies offers excellent opportunities to implement scientific findings in the practice of the people's nutrition. Exemplary supply of food in companies is the most effective nutrition education and propaganda [as it] enable[s] the workers to communicate practical experiences of rational feeding to their families.[22]

How important and how political this agency was is mirrored by the fact that even the basic structure of the menus was regulated by state laws which were published in the law gazette. The Institute for Nutrition in Potsdam undertook important tasks here, as it analysed and optimized the menus thoroughly and worked out a set of compulsory recipes for all East-German feeding facilities.[23] High participation figures show that the system was widely accepted (Table 15.2). But it was complicated and costly, as the state subsidized the meals heavily with up to 54 per cent of the cost of the raw materials.

Table 15.2   Participation in Catering 1964–1988

Target Group	Institutions	Participation (per cent)	
		1964	**1988**
		(%)	**(%)**
Infants and toddlers	Nurseries	16	72
Pre-school children	Kindergartens	49	95
Schoolchildren, apprentices	Schools	26	86
Students		20	57
Workers and Employees	Enterprises/Institutions	25	77
Normal and two-shift-system		20	72
Night shift		20	75

Source: *Ernährungsbericht* 1996, 68.

Another means of encouraging healthy eating habits was the production of health and diet foods. Already in 1953 a decree of the *Ministerrat der DDR* had ordered the production of a beer for diabetics[24] and in 1955 a directive for the food industry asked for the development of diet foods for diabetics and children,

---

22   Ludwig 1973: 70.
23   Zobel and Wnuk 1981.
24   'Die neuen Spezialbiere' 1954.

among them new and high-quality margarines.[25] Under the direction of the Centre for Diabetes and Metabolic Disorders in Berlin, the 'Association for Trademarks of Diet Foods' (*Warenzeichenverband Diätetische Lebensmittel*) was founded in 1969. Four enterprises from the food industry specialized completely in diet foods, five others had special departments for their particular products. As every single new food had to be examined and licensed by the state, bureaucracy slowed down the development.[26] In January 1970 a sales network was established: a store for diet foods was erected in every one of the 13 areas (*Bezirke*) of the GDR in order to distribute the products equally over the whole republic. Other production measures took up this blending of economic and health policies as, for example, by the development of a special breed of poultry. These 'broilers' were raised in a highly rational manner; their marketing included the propagation of health aspects as well as the establishment of special 'Broiler Restaurants'.[27] Moreover, the Five-Year-Plans for 1970–75 and 1976–80 asked for a reduction of sugar content in fruit products and an increase in the production of fat-reduced tinned fish and ready-to-eat meals. Additionally the latter plan stressed the provision of low-calorie bakery products, soft drinks and alcohol-free beer. Nevertheless, it took until 1978 for general regulations for diet foods to be released.[28]

These measures with regard to food consumption were backed up with educational efforts, which were targeted at the whole population. It was the German Museum of Hygiene (*Deutsches Hygiene-Museum*), which held a most important position in this process.[29] Throughout the GDR's history, it organized exhibitions, instructed teachers, produced and disseminated teaching materials. Since 1967 it had published its own popular Journal called *Deine Gesundheit* (Your Health). Being combined with its own Institute for Health Education, which collected scientific knowledge from other institutions, its research was directed towards the ways and means of transmitting health knowledge to the people. With regard to these questions it cooperated fully with the Institute for Nutrition in Potsdam, as well as with other institutions of education such as the 'Urania' in Berlin and with the National Committee for Health Education.[30] Campaigns began in the 1950s. Although they did not have the expected results, these efforts were not questioned but were reinforced in the 1970s, when new programmes on nutritional education were inaugurated, as for example the programmes on 'Diet and Performance' in 1972.

The starting point for all measures was the socialist personality and, at the Fourth Health Conference in 1972, the Ministry of Health declared that the true

---

25 Ministerium für Lebensmittelindustrie: *Direktive für die Lebensmittelindustrie im zweiten Fünfjahrplan*, Bundesarchiv Berlin, DE 1/25089: 10–13.

26 Schmidt 1971: 26.

27 Poutrus 2002.

28 *Gesunde Ernährung*, Part 1: 17–20.

29 Vogel 2003.

30 Fiedler 1973; Hirsch [1986].

socialist would be driven by a desire to maintain and to increase his capability and working capacity. Government and health administration equated health with physical capacity and with being a useful member of society and they expected people to adjust their eating habits to scientific findings.[31] By doing so, a biological model of eating was advanced once again, with a strong sense of rationality based upon the idea of one right, socialist, way of living. As before, economic means were used to change the situation, whereas the people were in fact left alone with their choice between a pickled knuckle of pork (*Eisbein*) and a butter cream cake.[32] Until the late 1970s and early 1980s the GDR simply used the average nutrient intake of different groups of society which were then compared to their physical needs.[33] A psychology of food hardly existed at all. There was no understanding of the fact that people did not eat rationally, and certainly not according to the principles of science. The state and scientists remained stuck in an old, ineffective mindset in their fight against obesity and diabetes.[34]

Health Policy Western Style

At the very beginning of the postwar era, government institutions in West Germany were eager to collect data on the actual nutritional situation: Food Commissions were established, teams were sent out to weigh people in the street, reports were requested from local health offices, in order to get some information on the nutritional status of the population with a view to maintaining physical strength and, thus, industrial productivity.[35] But as soon as the situation improved, these efforts were given up. The Allies thought that the strict authoritarian and central organization of the German Reich had been the major reason for the success of the National Socialist party; therefore they insisted on a decentralized, federal organization of scientific research and even permanent communal catering facilities were rejected.[36] Instead the family was regarded as the place to give birth to children, to raise and to feed them.[37] Due to this conservative family model, West Germany stuck to the old, half-time school system with children going home for lunch. Nevertheless education held an important position in the consumption sphere. Therefore institutions for the instruction of women in home economics were founded and a Federal Committee for Economic Instruction (*Bundesausschuss für volkswirtschaftliche Aufklärung*) was created with the help of money from

---

31   Sefrin 1972.
32   Sefrin 1972: 25.
33   Especially the works of Hans-Karl Gräfe. See Gräfe 1964.
34   Haenel 1972.
35   Sons 1983, Dinter 1999.
36   There were some exceptions to this as in case of the former Reich Research Institutes (Reichsanstalten), which were taken over by the state.
37   See Article 6 of the constitution.

the European recovery funds.[38] Although the Americans stressed the need to tie these institutes to a university or an institution of higher education, universities and scientists were rather reluctant to do so. Strongly orientated towards basic research, the natural scientists denied the value of teaching nutritional sciences and household economics. When a Federal Research Institute for Household Economics was eventually founded in 1952, its director, Elfriede Stübler, could get little cooperation and the institute's work was hindered. When – again on the initiative of the Americans – a first chair for nutritional science and household economics was founded in Giessen in 1960, its curriculum was primarily orientated towards biochemistry instead of social sciences. There was no comprehensive plan for the nutritional sciences at all. Instead, nutritional research was scattered over a number of universities and state institutes; it was also conducted in industry; and as far as metabolism and its disorders were concerned, in the university hospitals. The only overall institution was the German Society for Nutrition – the *Deutsche Gesellschaft für Ernährung* (DGE) – which was founded in 1954 according to the American model after a group of experts had visited the USA. This organization conducted none of its own research but was seen as a body making dietary recommendations and transmitting nutritional knowledge to the people.[39] It calculated food recommended nutrient allowances, gave nutritional advice and disseminated it by leaflets and brochures on the one hand and by training educators and advisers on the other.[40] Although the DGE claimed to be a non-governmental organization, there were people who would 'have an allergic reaction to official advice and consider the opposite to be true',[41] since its work was totally dependent upon subsidies from the state.[42]

During the following years the DGE set up advice centres in 35 towns all over West Germany and in 1957 an 'Institute for Nutritional and Dietetic Advice' at the Medical Academy in Düsseldorf. In 1967 a research institute and clinic for Diabetes research was erected there, in which 'Nutrition Advisers' (*Ernährungsberater*) were educated.[43] As in the East, it was believed that the consumer would decide on rational grounds and thus eat in the recommended, healthy way, which was taught in an authoritarian manner. In fact, nutrition was regarded to be an entirely private affair. This opinion hindered the onset of a concerted action against diabetes, although the Insulin Committee, which had been functioning from 1923 onwards, was re-established in 1949. The publication of central recommendations by the Committee in 1949 was criticized heavily. Physicians claimed to be the only

---

38   Thoms 2004a, Richarz 1991.

39   Melzer 2003.

40   Janssen 2001.

41   There were many aspects of the new association reminiscent of National Socialist nutrition policy.

42   See the letter from 12.3.1963, Bundesarchiv Koblenz, B 142, No. 2494; Thoms 2004b.

43   Berger, Lilla and Petzoldt 1990.

competent persons to advise diabetics individually in a doctor's practice. Neither diabetes centres nor a diabetics register were appropriate. As a result, there was no systematic survey of the number of diabetics or the development of their disease, so that there were no consistent data.[44] Where diabetes information centres existed, they were attached to special (university) clinics and followed the interest of the respective directors, rather than general lines of advice.

Consequently several scientific, as well as self-help, associations took over the task of informing the public and the sick. As early as 1951 the Diabetes League (*Diabetiker Bund*) was founded, and the German Diabetes Association (*Deutsche Diabetes Gesellschaft*) followed in 1964. By contrast with East Germany, the question of diet foods was left to the free market: There was no policy of licensing diet foods. Instead the Association of the Dietetic Food Industry (*Verband der diätetischen Lebensmittelindustrie*) set standards which gave guidelines for the production and advertisement of diet foods and became the basis for a decree on diet foods from June 1963 (*Verordnung über diätetische Lebensmittel*).[45]

This supremacy of economic factors and the free market meant that responsibilities for food questions were concentrated on the production side on behalf of the Ministry of Agriculture. Consequently the DGE was first sponsored by this ministry and only from 1969 onwards by the Ministry of Health. The central and well-organized health system of the GDR, with the emphasis on prevention, was rejected during the 1960s, as public controversies show. All central planning was refused, because it was conceived as the system of the political enemy. It was said that this mode of health care would be incompatible with freedom and dignity and would thus be incompatible with democracy, especially if it put the common welfare above the individual's rights and wellbeing.[46] Although the doctors complained about the increasing body weight of their patients and about the lack of epidemiological data, it took until 1984 before the very first representative food consumption study was undertaken. In 1984 the Vera study started in the East, whilst the National Consumption Study was started in the West in 1985 in order to get more reliable and differentiated statistical data on the individual nutritional situation in Germany. Before this, on both sides of the Wall, food consumption had only been calculated from production figures.

The Outcomes of these Policies in West and East Germany

Due to a complex mix of economic and political factors, the different histories of the two German states were accompanied throughout by differences in food consumption. East Germany consumed more fat, sugar, grain, potatoes and

---

44   Schliack 1959.
45   Holthöfer, Juckenack and Nüse 1966: 422ff.
46   Norpoth, Leo, 'Die geplante Gesundheit', in: *Rheinischer Merkur*, 17 August 1962.

vegetables, but less fruit and cheese, cocoa and coffee.[47] Despite all the differences with regard to the economic, political and health care systems, the energy balance, as well as the incidence of obesity, was almost the same in both parts of Germany shortly before reunification in 1989 (see Figure 15.2).[48] Seemingly neither system had managed to control the problem. The human factor withstood the measures of central control in the East as well as the more rational appeal to the individual in the West.

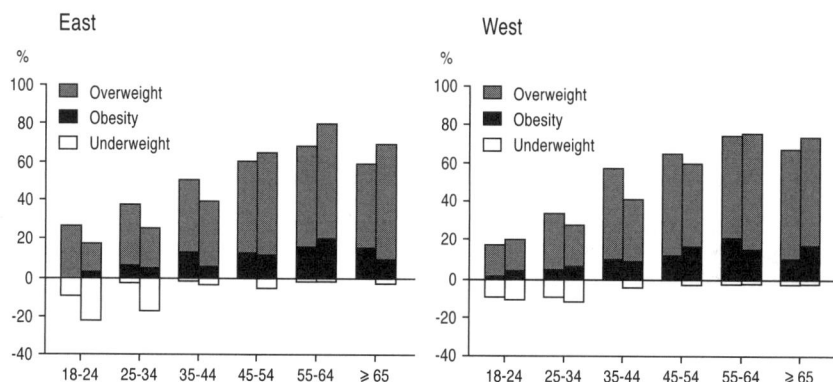

Figure 15.2 **Percentage of Obese Persons in the Population According to Age Groups in East and West Germany, 1987/88**

Source: Redrawn after *Ernährungsbericht* 1992, 34.

On both sides of the Berlin Wall the process of rethinking nutrition education began only in the late 1980s and had to be addressed carefully by stressing the dominance of the natural over the social sciences.[49] Only since then has nutrition policy begun to emphasize positive elements, such as health, beauty, mobility, self-consciousness and a more positive health aim instead of criticizing 'dangerous' behaviour and painting pictures of horror with regard to the risks.[50] This demonstrates, firstly, that in case of Germany the effects of different political systems and different systems of health education should not be overestimated and, secondly, that simple cause-effect models are inappropriate in nutrition affairs. Only recently has a discussion on the effectiveness of the East German health-care system begun anew in Germany because the current Federal German health-care system is in difficulties.[51] In the face of the system's financial and demographic

---

47 DGE, *Ernährungsbericht* 1992: 26, Kutsch and Weggemann 1996.
48 DGE, *Ernährungsbericht* 1992: 21–7.
49 Ketz and Hanke 1981: 11–38.
50 Ketz and Hanke 1981: 66.
51 See the proposals in Bruns et al. 2004.

problems, it seems worthwhile to reconsider the East German centralized system, leaving behind all ideological objections, as it is much cheaper than the highly individualized West German model, which obviously did not result in a better nutritional status for the people. Both systems led to similar health risks and nutrition-related epidemiological patterns on either side of the Wall. It may be reasonable, therefore, to try at least some elements of the former Eastern model.

References

'Die neuen Spezialbiere unserer Brauindustrie' [The New Beers of our Brewing Industry], in *Die Lebensmittelindustrie* 1, 1954, 13.

Beller, A.S. *Fat and Thin: A Natural History of Obesity*, New York 1977.

Berger, M., Lilla, M. and Petzoldt, R. *25 Jahre Deutsche Diabetes-Gesellschaft* [25 years of the German Diabetes Association] Mainz, 1990.

Bertram, F. 'Zur Pathogenese der Regulationskrankheiten' [Pathogenesis of Regulation Diseases], *Deutsche Medizinische Wochenschrift* 75, 1950, 97–101.

Boldorf, M. *Sozialfürsorge in der SBZ/DDR 1945–1953. Ursachen, Ausmaß und Bewältigung der Nachkriegsarmut* [Social Care in the GDR 1945–1953. Causes, Extent and Coping with Poverty after War], Stuttgart, 1998, 80.

Buns, W., Menzel, R., Panzram, G. and Seige, K. *Die Entwicklung der Diabetologie im Osten Deutschlands von 1945 bis zur Wiedervereinigung* [The Development of Diabetology in East Germany from 1945 to Re-Unification] Hildesheim, 2004.

Damm, G. 'Zur Ätiologie der Fettsucht' [On the Aetiology of Obesity], *Deutsche Medizinische Wochenschrift* 76, 1951, 1187–9.

Dinter, A. *Berlin in Trümmern. Ernährungslage und medizinische Versorgung der Bevölkerung Berlins nach dem II. Weltkrieg* [Berlin Wrecked: the Food Situation and Health Care in Berlin after WW II], Berlin 1999.

Dittrich, M. *Die Rolle der wissenschaftlichen Schule von Gerhardt Katsch (1887–1961) für die Entwicklung der Diabetesforschung* [The Scientific School of Gerhardt Katsch and its Role in the Development of Diabetes Research], Diss. B, Ms. 2 vols, Greifwald, 1986.

Elkeles, T., Niehoff, J-U., Rosenbrock, R. and Schneider, F. (eds), *Prävention und Prophylaxe. Theorie und Praxis eines gesundheitspolitischen Grundmotivs in zwei deutschen Staaten 1949–1990* [Prevention and Prophylaxis: Theory and Practice of a Health-Related Issue in Two German States, 1949–1990], Berlin, 1991, 51–74.

Ernährungsbericht 1992 [Nutrition Report], ed. by the *Deutsche Gesellschaft für Ernährung*, Frankfurt a.M., 1992.

Fiedler, M. 'Zur Effektivität medizinischer Veranstaltungen im Berliner Vortragszentrum der URANIA' [On the Effectiveness of Medical Activities of the URANIA], *Sozialistische Ideologie und Gesundheitserziehung*.

Materialien des 2. Symposium des Instituts für Gesundheitserziehung im Deutschen Hygiene-Museum in der DDR vom 4.–6. April 1973 in Dresden, Dresden o.J., [1973], 292–6.

Gilman, S.L. *Fat: A Cultural History,* Cambridge, 2008.

Gräfe, K.-H. 'Entwicklung des Fettverbrauchs im letzten Jahrzehnt, unter besonderer Berücksichtigung der Verhältnisse in der DDR' [Development of Fat Consumption During the Last 10 Years], *Die Nahrung* 8, 1964, 269–82.

Gries, R. *Die Rationen-Gesellschaft. Versorgungskampf und Vergleichsmentalität: Leipzig, München und Köln nach dem Kriege* [The Rationed Society], Münster, 1991.

Grosse-Brockhoff, F. 'Die Bedeutung der Adipositas als Krankheitsursache, ihre Therapie und Prophylaxe' [The Importance of Obesity as a Cause of Disease], *Deutsche Medizinische Wochenschrift* 78, 1953, 399–402, 435–9.

Grundy, S.M., Brewer, H.B., Cleeman, J.I., et al. 'Report of the National Heart, Lung, and Blood Institute/American Heart Association Conference on Scientific Issues Related to Definition', *Circulation* 109, 2004, 433–8.

Hacking, I. *The Taming of Chance,* Cambridge, 1990.

Haenel, H. 'Ernährungssituation und Ernährungsprognose' [Nutritional Situation and Nutrition Prospects], *Ernährungsforschung* 18, 1972, 29–43.

Hauner H., Köster I., Schubert I. 'Trends in der Prävalenz und ambulanten Versorgung von Menschen mit Diabetes mellitus' [Trends in Prevalence and Outpatient Care for People with Diabetes Mellitus], *Deutsches Ärzteblatt* 104, 41, 2007, 2799–805.

Hirsch, W. 'Ernährungsaufklärung in der Arbeit der Urania' [Educational Advertising in Nutrition in the Work of the URANIA], *Ernährung – Gesundheit – Genuß. Praxis und Wissenschaft,* o.O. [Berlin] o.J. [1986], 80–82.

Holthöfer, H. and Juckenack, A. *Deutsches Lebensmittelrecht* [German Food Law], *Bd. II: Neben dem Lebensmittelgesetz geltendes Lebensmittelrecht, 4. Aufl.,* Berlin, 1966, 422–544.

Janssen, P. *Zur Geschichte der Ernährungsberatung in der Bundesrepublik Deutschland. Die Deutsche Gesellschaft für Ernährung e.V.* [On the history of nutritional advice in Germany. The German Society for Nutrition], Giessen, 2001, 11–13.

Ketz, H.-A. and Hanke, P. 'Probleme und Entwicklungsrichtungen der gesellschaftlichen Grundlagen der Ernährung' [Problems and Directions in the Development of Nutrition], in Ketz, H.-A. and Hanke, P. (eds) *Ernährung – gesellschaftlich bedingt,* Berlin, 1981, 11–38.

Kutsch, T. and Weggemann, S. (eds), *Ernährung in Deutschland nach der Wende: Veränderungen in Haushalt, beruf und Gemeinschaftsverpflegung* [Diet in Germany after the Political Turn: Changes in Household, Occupation and Catering], Bonn 1996.

Linow, F., Lewerenz, H-J. and Möhr, M. 'Zur Geschichte des Institutes für Ernährungsforschung in Potsdam-Rehbrücke. Die Entwicklung von 1946 bis 1950' [On the History of the Institute for Nutrition Research in Potsdam: Developments from 1946 to 1951], *Ernährungsforschung* 41, 1996, 1–25.

Ludwig, W. *Grundriß der Gesundheitserziehung. Wissenschaftliche Grundlagen und Systematik* [Outline of Health Education: Scientific Basics and Systematics], Berlin 1973, 70.

Meier, E. 'Über die Variation von Körpergrößen und Körpergewichten bei Erwachsenen nach Untersuchungen an der Berliner Bevölkerung' [On the Variation of Height and Body Weight in Adults According to Examinations of the Berlin Population], *Archiv für Hygiene* 140, 1956, 304–6.

Melzer, J. *Vollwerternährung. Diätetik, Naturheilkunde, Nationalsozialismus, sozialer Anspruch* [Whole Value Nutrition: Dietetics, Naturopathy and Social Claims], Stuttgart, 2003.

Niehoff, J-U. and Schrader, R-R. 'Gesundheitsleitbilder – Absichten und Realitäten in der Deutschen Demokratischen Republik' [Health Concepts – Intentions and Realities in the GDR], in Elkeles, T., Niehoff, J-U., Rosenbrock, R. and Schneider, F. (eds), *Prävention und Prophylaxe. Theorie und Praxis eines gesundheitspolitischen Grundmotivs in zwei deutschen Staaten 1949–1990*, Berlin, 1991, 51–74.

Porter, M. *The Rise of Statistical Thinking, 1820–1900,* Princeton, 1988.

Poutrus, P.G. *Die Erfindung des Goldbroilers. Über den Zusammenhang zwischen Herrschaftssicherung und Konsumentwicklung in der DDR* [The Invention of the Gold Broiler: on the Relationship Between Consumer Protection and Regulation in the GDR], Köln, 2002.

Richarz, I.O., *Haus und Haushalten. Ursprung und Geschichte der Haushaltsökonomik [House and Households: Rise and History of Household Economics]*, Göttingen 1991.

Schagen, U. and Schleiermacher, S, '100 Jahre soziale Medizin in Deutschland' [100 years of Social Medicine in Germany], *Das Gesundheitswesen* 68, 2006, 81–93.

Scheunert, A. *Ernährungsprobleme der Gegenwart* [Present Nutritional Problems], *Sitzungsberichte der Akademie der Landwirtschaftswissenschaften zu Berlin,* 1, 1952.

Schliack, V. 'Diabetes-Reihenuntersuchungen' [Surveys in Diabetes], *Deutsche Medizinische Wochenschrift* 84, 1959, 1446–8.

Schmidt, F. 'Die Gestaltung der Kooperation bei der schrittweisen Herausbildung eines Systems der gesunden Ernährung in der DDR und die Aufgaben dazu im Volkswirtschaftsplan 1971 sowie im Perspektivplanzeitraum bis 1975' [The Formation of Cooperation in the Step-by-Step Evolution of a System of Healthy eating in the GDR], in *Die Kooperation bei der Herausbildung eines Systems der gesunden Ernährung in der DDR*, Berlin, 1971, 26.

Schmidt-Semisch, H. (ed.), *Kreuzzug gegen Fette. Sozialwissenschaftliche Aspekte des gesellschaftlichen Umgangs mit Übergewicht und Adipositas*, Wiesbaden, 2008.

Sefrin, M. 'Die Rolle der Ernährung beim Gesundheitsschutz der Bevölkerung' [The Role of Nutrition in the Health Protection of the Population], *Ernährungsforschung* 17, 1972, 13–28.

Semmler, L. and Ketz, H-A. *Gesunde Ernährung* [Healthy Eating], Dresden.

Sons, H.U. *Gesundheitspolitik während der Besatzungszeit. Das öffentliche Gesundheitswesen in Nordrhein-Westfalen 1945–1949* [Health Policy during Occupation: the Public Health System in North Rhine Westphalia 1945–49], Wuppertal, 1983.

Stöckel, S. (ed.) *Prävention im 20. Jahrhundert. Historische Grundlagen und aktuelle Entwicklungen in Deutschland* [Prevention in the 20th Century. Historical Background and Current Developments in Germany], Weinheim, 2002.

Thoms, U. 'Learning from America? The Travels of German Nutritional Scientists to the USA in the Context of the Technical Assistance Program of the Mutual Security Agency and its Consequences for West German Nutritional Policy', *Food & History* 2, 2004a, 117–52.

Thoms, U. 'Die "Hunger-Generation" als Ernährungswissenschaftler 1933–1964. Soziokulturelle Gemeinsamkeiten oder Instrumentalisierung von Erfahrung?' [The 'Hunger-Generation' as Nutritional Scientists 1933–64: Sociocultural Common Grounds or Instrumentalization of Experience?], in Middell, M, Uekötter, F. and Thoms, U. (eds), *Wie konstruiert die Wissenschaftsgeschichte ihre Objekte?*, Leipzig 2004b, 133–53.

Thoms, U. 'Vitaminfragen – kein Vitaminrummel? Die deutsche Vitaminforschung in der ersten Hälfte des 20. Jahrhunderts und ihr Verhältnis zur Öffentlichkeit' [Vitamin Question – no Vitamin Whoopee? German Vitamin Research in the First Half of the 20th Century and its Relation to Society], in Nikolow, S. and Schirrmacher, A. (eds), *Wissenschaft und Öffentlichkeit als Ressource füreinander*, Bielefeld, 2007.

Thoms, U. 'Körperstereotype. Veränderungen in der Bewertung von Schlankheit und Fettleibigkeit in den letzten 200 Jahren' [Body Stereotypes: Changes in the Valuation of Thinness and Obesity in the last 200 Years], in Wischermann, C. and Haas, S. (ed.), *Körper mit Geschichte. Der menschliche Körper als Ort der Selbst- und Weltdeutung*, Stuttgart, 2000, 281–308.

Troschke, J. von 'Organisation und Praxis der Prävention in der Bundesrepublik Deutschland' [Organization and Practice of Prevention in the FRG], in Elkeles, T., Niehoff, J-U., Rosenbrock, R. and Schneider, F. (eds), *Prävention und Prophylaxe. Theorie und Praxis eines gesundheitspolitischen Grundmotivs in zwei deutschen Staaten 1949–1990*, Berlin, 1991, 75–106.

Ulbricht, W. 'Das Programm des Sozialismus und die geschichtliche Aufgabe der SED' [The programm of socialism and the historical task of the SED], *Protokoll der Verhandlungen des VI. Parteitages der Sozialistischen Einheitspartei Deutschlands, 18–21 January 1963 in der Werner-Seelenbinder-Halle zu Berlin*, vol. 1, Berlin, 1963, 1–3.

Verfassung der Deutschen Demokratischen Republik vom 6. April 1968 [Constitution of the German Democratic Republic], Berlin 1981.

Vogel, M. *Das Deutsche Hygiene-Museum Dresden: 1911–1990* [The History of the German Museum of Hygiene in Dresden], Dresden, 2003.

Voß, P. 'Wie erreichen wir den Bürger? Überlegungen aus der Sicht des Deutschen Hygiene-Museums in der DDR' [How to Get Through to the People: Thoughts from the Perspective of the German Museum of Hygiene], in *Ernährung – Gesundheit – Genuss – Praxis und Wissenschaft*, Berlin, [1986], 64–7.

World Health Organization 'Obesity: Preventing and Managing the Global Epidemic: Report of a WHO Consultation', *WHO Technical Report Series* 894, Geneva, 2000.

Zobel, M. and Wnuck, F. *Neuzeitliche Gemeinschaftsverpflegung* [Modern Collective Feeding], vol. 1, 12th ed., Leipzig, 1981.

Chapter 16
Conclusion

Derek J. Oddy and Peter J. Atkins

When the contributors to ICREFH's tenth symposium assembled in Oslo in September 2007, they met to discuss what, for them, was an unusual research question. Never before had they attempted to consider changes in food consumption over such a short time span – a single century – which assumed that all the countries under consideration experienced similar stages of development within the same period. The question posed was, in essence, how did food consumption change in Europe in the twentieth century and why are so many Europeans now overweight? The outcome of the discussion summarized in these chapters shows a considerable similarity of approach and invites some attempt at comparison by the editors. The criticisms frequently levelled by reviewers at volumes emanating from conferences, namely that there is a lack of coherence in the mix of topics, is hardly valid in this book. Its sections consider the major influences on European food consumption in the twentieth century and invite comparisons across national boundaries. In short, all European countries have been affected by the industrialization of their food industries and, by the end of the twentieth century, every state in Europe had begun to show concern for the extent to which its population had become overweight and to consider the future implications if this weight gain were unchecked.

This volume has assembled evidence from nine European countries which were at different stages of economic development as the twentieth century began and which have followed different paths towards industrialization and economic maturity. They are listed in Table 16.1. Not all were nation states when the century commenced. Only one of the nine was neutral during both World Wars, and only two avoided enemy occupation; however the majority experienced traumatic upheavals leading to loss of life, especially amongst males in the 20–35 age groups, as well as civilian casualties. Continental Europe was a focus for international aid after both World Wars. The Save the Children Fund was formed in London in 1919 to send relief to Europe following the 1914–1918 wartime blockade of Germany and Austria-Hungary. As the principal contributor to the International Save the Children Fund founded in 1920, the British Save the Children Fund also helped in the feeding of famine victims in the Volga region of Russia during the 1921–1922 famine there.[1] During the Second World War the United Nations

---

1 In August 1921, the International Red Cross in Geneva set up the International Committee for Russian Relief (ICCR) with Dr Fridtjof Nansen as High Commissioner. The main participants were the USA's American Relief Administration (ARA) and the

Relief and Rehabilitation Administration (UNRRA) was set up in 1943 to carry out an aid programme to liberated areas of Europe, principally by providing food. UNRRA did this from 1944 to 1947, as mentioned in Chapters 3 and 4.[2] Until the mid-twentieth century, therefore, recovery programmes assumed sizable areas of Europe contained undernourished populations. From 1949 to the 1960s, energy and nutrient requirements were debated by nutritionists and physiologists through international committees seeking to establish minimum standards, which applied in Europe as much as in those underdeveloped areas of Africa and Asia coming to be known as the 'Third World'.[3] To make international comparisons of a scientific kind required the concept of average physical development – 'reference man' and 'reference woman' – whose body weight was put at 65 kg and 55 kg respectively and whom, if young adult males, would have 12.5 per cent and if young adult females 25 per cent of energy reserves as body fat, or adipose tissue. Between 1957 and 1967, a series of reports from the United Nations agencies, Food and Agriculture Organization (FAO), and FAO jointly with the World Health Organization (WHO), produced standard requirements of energy and nutrients for work, leisure and rest.[4]

By the 1960s there was a growing recognition that populations across Europe were progressively gaining weight. This progression was most noticeable in older age groups, a trend already noted in the USA by the life insurance companies and which, in Europe, had been initiated by the post-rationing surge in eating. Monitoring weight gain has increased during the last thirty years notably by the WHO MONICA Project and the WHO SuRF Reports, as well as the European Commission.[5]

British Save the Children Fund. Fundraising in Britain anticipated modern emergency relief operations by using full-page newspaper advertisements and collections in cinemas based on a fundraising film of the famine area. The first feeding centre was opened in October 1921 in Saratov. Some ten million people were fed by the ICCR of which the ARA provided the overwhelming amount of the funding.

2   UNRRA pre-dated the formation of the United Nations. Originally intended to help refugees and displaced persons in Europe, the chief beneficiaries were Albania, Austria, Byelorussia, Czechoslovakia, Greece, Italy, Poland, the Ukraine, and Yugoslavia.

3   The economist and demographer, Alfred Sauvy, in an article published in the French magazine *L'Observateur* 14 August 1952, coined the term *Third World* to refer to countries that were unaligned with either the Communist Soviet bloc or the North Atlantic Treaty Organization bloc during the Cold War (1945–89).

4   FAO Committee on Calorie Requirements 1950, FAO Nutritional Studies No. 5, Second FAO Committee on Calorie Requirements 1957, FAO Nutrition Studies No. 15, Protein Requirements Report of the FAO Committee 1957, FAO Nutritional Studies No.16, Protein Requirements. Report of a Joint FAO/WHO Expert Group 1965. The term 'reference man' has been used in the USA since the 1970s in connection with exposure to radiation and its original meaning has become obscured.

5   WHO *Multinational Monitoring of trends and determinants of Cardiovascular Disease*: 'The main hypothesis of the WHO MONICA Project is to assess whether 10-year

Table 16.1   Demographic Data for Selected European Countries, 1965–70

Country	Estimated population 1969 (M)	Birth rate (/000)	Death rate (/000)	Gross national product/ head (US$)
Austria	7.4	17.4	13.0	1,150
Czechoslovakia	14.4	15.1	10.1	1,010
France	50.0	16.9	10.9	1,730
East Germany	16.0	14.8	13.2	1,220
West Germany	58.1	17.3	11.2	1,700
Norway	3.8	18.0	9.2	1,710
Soviet Union	241.0	18.0	8.0	890
Spain	32.7	21.1	8.7	640
United Kingdom	55.7	17.5	11.2	1,620
Yugoslavia	20.4	19.5	8.7	510
Europe	456.0	18.0	10.0	1,230

Source: Davidson et al. 1975, Table 51.1.

Unfortunately, for comparative purposes no countries collected aggregate statistics that showed physical stature as well as energy intakes. In the middle of the twentieth century few countries could make even general statements about trends in weight gain by their populations. Nevertheless, it is clear that the idea of making any estimate of body weights, however limited or general, did not occur before the 1960s. The national case studies in this volume give only a sketchy outline of the body-weight trend in Europe. In the Austrian Tyrol the suggestion made is that the population in the 1960s was still characterized by weight deficits rather than excess weight, though by contrast its neighbour, Czechoslovakia, categorized its adult male population as 25 per cent obese and its female population as over 50 per cent obese. Czechoslovakia, however, seems to have applied over-rigorous standards based on the Broca Index, so that these percentages at the mid-century substantially exceed its current levels based on the Body Mass Index. The application of differing standards makes comparison almost impossible.

---

trends in incidence and mortality from cardiovascular disease are related to changes in known risk factors. Weight (relative weight, degree of overweight, obesity etc.) was not originally included as one of these risk factors although data on weight and height have been collected from the beginning of the MONICA survey periods. Although the matter is still somewhat under debate, overweight is now considered as one of the risk factors for the main hypothesis.' See the MONICA website: http://www.ktl.fi/monica/, and WHO 2005.

In the United Kingdom, following the FAO Second Committee report in 1957, individuals with excessive levels of 'energy reserves' or body fat were classified as obese if they amounted to more than 20 per cent of body weight for young adult males and more than 30 per cent for young females. However, no standard method of assessment was envisaged: nutritionists in the United Kingdom relied principally on the more accurate skin fold measurement to assess amounts of body fat. The Quetelet, or Body Mass, Index was not used.

The fat content of the diet began to attract notice as the incidence of cardiovascular disease increased during the 1950s.[6] During the next two decades, the beginnings of a secular trend in weight gain were observable and the Body Mass Index became generally accepted as a standard measure of weight variation for individuals. In Norway, the weight of men aged 40 years increased by 9 kg (19 lbs) from the 1970s until the end of the century while the bodyweight of women of the same age increased by 4 kg (9 lbs). In Soviet Russia, it was estimated that 8 per cent of the population were obese by the 1980s, a figure not dissimilar from the United Kingdom's assessment that obesity was rising from 6 per cent in that decade, as shown in Table 16.2, By the end of the twentieth century, further progressive rises were postulated: 6 per cent were said to be obese in Norway; around 13 per cent in Russia, 22 per cent of men and 23 per cent of women in the United Kingdom but fewer than 9 per cent in Austria.

Table 16.2   Percentage of Population Aged 16–64 Defined as Obese in England

Date	Men (%)	Women (%)
1966	1.2	1.8
1972	1.7	1.7
1982	6.2	6.9
1989	10.6	14.0
1999	18.7	21.1
2007	23.6	24.4

Sources: Comptroller and Auditor General 2001; WHO, *Global Database on Body Mass Index*; *UK National Statistics*

This progressive but uneven rise in body weights can be extended to cover all European Union (EU) countries. Data supplied which cover the years 1996 to 2003 give an EU average of 47.5 per cent of overweight and obese adults aged

---

6   'Dietary Fat and Cardio-Vascular Disease', *British Food Journal* January 1962: 1–2.

15 years and over, of whom 34.1 per cent are overweight and 13.4 per cent obese. Data for selected countries are shown in Table 16.3. Although Table 16.3 provides only a partial coverage of European countries, it does include the two extremes of the range: Norway with the lowest incidence, having 31.5 per cent of overweight and obese and the United Kingdom with the highest at 61 per cent.

Table 16.3   Prevalence of Overweight and Obesity Amongst Adults in Selected European Countries, 2004

	Overweight			Obese		
	Male	**Female**	**All**	**Male**	**Female**	**All**
	(%)	**(%)**	**(%)**	**(%)**	**(%)**	**(%)**
Austria	37.9	25.6	–	18.5	15.6	–
Czech Rep	42.8	30.4	36.4	13.5	15.3	14.4
France	35.1	21.2	27.8	9.4	9.2	9.3
Germany	48.0	31.3	39.4	18.8	21.7	20.3
Norway	31.4	19.6	25.4	6.4	5.9	6.1
Russia	30.7	27.4	28.9	10.3	21.6	16.0
Slovenia	43.3	29.7	36.2	12.6	12.0	12.3
Spain	44.1	27.8	35.7	13.0	13.5	13.3
UK	43.9	33.6	38.3	22.3	23.0	22.7
EU27	42.8	29.5	–	16.2	18.1	–

Source: European Commission, *Eurostat*.

Notes: (a) Adults aged 15 years and over; (b) Data from National Health Interview Survey 1996–2003; (c) Austrian, Russian and EU27 data from International Obesity Taskforce.

Some care is needed with Table 16.3 because it lists only those countries discussed in the present book. Beyond this selection there are some remarkable variations. According to the International Obesity Taskforce, the peak of male obesity lies in Malta, with 24.3 per cent, and the lowest prevalence is in Sweden at 7.0 per cent.[7] For women, the extremes are in England (23.2 per cent and a long way ahead of nearest rivals Greece and Lithuania) and Sweden (7.5 per cent). The situation is little different if one combines overweight (BMI 25–29.9) with obesity (BMI ≥30),

---

7   It is noticeable that percentages for particular countries vary greatly between the WHO, EU and IOTF databases and this is due to the sampling techniques employed, the definition of overweight and obesity, and the age groups included or techniques of age standardization used.

although now the Maltese (70.1 per cent) and English (65.8 per cent) men are more closely challenged by Finland (63.2 per cent), Germany (61.7 per cent) and Greece (61.0 per cent), and the English women (54.3 per cent) are only marginally ahead of Lithuania (53.3 per cent) and Portugal (52.1 per cent). At the low end are two astonishing figures: Swedish men (32.9 per cent), who are more than ten points below that of any of the other EU countries, and Italian women (29.5 per cent), who are nearly 20 points below the EU27 average of 47.6 per cent.[8]

Obesity geographies are notoriously problematic in view of issues of comparability between national statistical services. But a map of European mean BMI is instructive. It does not show a clear correlation with economic development, or with any regional cultural variables. Yet there does seem to be a fragmented tendency in south eastern Europe for BMIs, both male and female, to be high and to match the levels found in the Middle East and north Africa.[9] More research, at the sub-national scale, is desirable if health messages are to be properly attuned to such variations and their historical roots.

What inferences can be drawn from the diverse figures mentioned above? Postwar Europe has achieved remarkable regularity of food supplies: in part, this has been an outcome of the EU's system of managed agriculture and state subsidies, even if food costs have been higher in the EU than elsewhere. Harvest fluctuations did not cause food consumption to oscillate in the second-half of the twentieth century. If food supplies have been more secure, life has also become more sedentary, both at work and in the home, and this reduction in energy expenditure has been accentuated by the widespread use of mechanized personal transport. Increasing bodyweights and expanding waistlines have been the result of excessive energy intakes, notably in the final thirty years of the twentieth century. In Section 3, several chapters have outlined an attack on energy reserves as fat. The attack on fat arose earliest in the interwar years amongst the middle classes – and amongst the young who were more sensitive to fashion – but made little impression on the working classes. With the increasing security of the food supply, eating to achieve good health as advocated in the late nineteenth and early twentieth centuries became outmoded. Only Chapter 5 reports food consumption as recently as the last decade of the twentieth century – which is appropriate, as the UK was becoming the most overweight nation in Europe. It was a significant outcome given the UK's exposure to the influence of Americanization and the extensive availability of industrialized food products supplied by major food manufacturing companies. This was the ultimate result, reflecting the industrialization of the diet which was becoming widespread across Europe by the end of the twentieth

---

8   See: http://www.iotf.org/database/documents/v2PDFforwebsiteEU27.pdf [accessed: 1 March 2009].

9   The mean BMI data are to be found in the *Surf2 Report*, WHO 2005. Convenient maps are published by the British Heart Foundation: http://www.heartstats.org/temp/ESspFigsp10.3aspweb08.pdf and http://www.heartstats.org/temp/ESspFigsp10.3bspweb08.pdf [both accessed: 20 April 2009].

century. Dishes and meals manufactured by food processors were notably high in sugar, fats and salt, which enabled pleasing flavours to be created at low cost. These foods attracted low-income families across Europe. They required little time or skill to remove from the refrigerator or freezer and heat up. They tended to be low in protein and lack any significant amounts of fruit and vegetables.

In recent years, much international emphasis has been placed on weight gain. The use of Body Mass Index has been adopted by WHO, which defines it as 'an interactive surveillance tool for monitoring nutrition transition'.[10] As it stands, that statement is grossly oversimplified. Changes in BMI are more complex than being due to the 'nutrition transition' even though Table 16.4 shows a general increase in energy intakes in the countries studied in this volume. Food historians have been recording the 'nutrition transition' since the 1960s which they have identified as a component of industrialization and urbanization. However, the trends they have documented have made little impact on the formation of public policies or public health management. What is important about the wider acceptance of BMI for food history is that it is confirmation of the extent to which food habits have changed in developed countries, and how changing energy expenditure at work and in the home, together with an increasingly sedentary lifestyle, have become significant causes of present-day health problems.

Table 16.4 Energy Intakes in Selected European Countries (kcal/person/day)

Country	1961	1970	1980	1990	2000
	(kcal/day)	(kcal/day)	(kcal/day)	(kcal/day)	(kcal/day)
Austria	3,190	3,232	3,353	3,485	3,767
Czech Republic	3,301[1]	3,357[1]	3,341[1]	3,649[1]	3,111
France	3,194	3,301	3,376	3,512	3,594
Germany	2,887	3,146	3,338	3,307	3,430
Norway	3,003	3,021	3,350	3,143	3,365
Russian Federation	3,095[2]	3,354[2]	3,369[2]	3,363[2]	2,919
Slovenia	3,048[3]	3,424[3]	3,662[3]	3,621[3]	3,055
Spain	2,631	2,732	3,062	3,246	3,362
United Kingdom	3,290	3,327	3,159	3,268	3,380

Source: Food and Agriculture Organization, FAOSTAT.

Notes: (1) Czechoslovakia; (2) USSR, (3) Yugoslavia.

---

10 WHO, *Global Database on Body Mass Index*.

While obesity was known in the nineteenth century, society's response to the condition was to regard it as a matter for the individual. Culturally, refined people were not obese. In as far as it was to be found amongst the upper and middle classes, it reflected new money from trade and industry rather than old money from land and political power. There were opportunities for obesity to be 'treated' by summer visits to spas across Europe from Bath to Bohemia to bathe, drink the waters, take exercise and eat abstemiously. Anyone able to afford it might find themselves mixing with the royal families of the European monarchies. The typical range of treatments has been fully detailed for ICREFH by Sabine Merta.[11] Dieting, abstinence and exercise implied a general recognition that individuals had something akin to a moral responsibility to restrain their appetites for food and drink. Nevertheless, Chapter 12 in this volume does present another view, also from the late nineteenth and early twentieth centuries, which hints at the early medicalization of treatments in France for being overweight or obese. This did not mean that physicians were eager to assess their patients by any physiological analysis derived from organic chemistry involving measuring intakes of carbon or nitrogen. Instead, much medical opinion ignored new advances in science and sought to discover non-nutritional causes of obesity. From this viewpoint a range of slimming regimes blossomed during the first forty years of the twentieth century, which began to produce dietary advice on an almost industrial scale. The proponents of various slimming regimes were proprietorial in their advice, as it was necessary to justify their consultation fees. This restricted their influence to the metropolitan middle and upper classes whose lifestyle was supported by unearned incomes. Herein lay the origins of a strand of dietary advice which lasted throughout the twentieth century. In short, it created an excuse-system by which the 'patient' might blame obesity or excessive body weight upon his or her endowment – such as heavy bones, large skeletons or unusual metabolism.

Wartime rationing, food shortages, and even famines in the first-half of the twentieth century, halted the progress of the slimming and dieting industry. By 1950, many people in Europe in the forties to sixties age groups, whose weight would have been rising with age before 1939, were instead physically worn and haggard. However, the surge of compensation eating which followed during the postwar economic recovery caused populations across Europe – particularly in the west and central regions – to begin putting in weight. Subsidized agriculture resulting from the Common Agriculture Policy (CAP) led to the over-supply of food in Europe.[12]

There was little opportunity in the early 1950s for the diet advisers to influence people who were enjoying freedom of consumption again. No one wanted to cut back on food. Fashion-conscious younger women looked to their foundation garments to enhance their silhouettes rather than restrict their food intakes. But

---

11   See Merta 2005.

12   The CAP was adopted in 1960 and came into force in 1962. The UK also subsidized food prices through a system of agricultural managed prices from 1947 to 1973.

people of the postwar era soon began to notice the way that 'diseases of affluence' were being discussed, not just in professional science journals but also in newspapers which introduced medical terms such as atherosclerosis and coronary heart disease in their news items. Initially blame fell on food habits such as the high level of sugar consumption – encouraged during the Second World War – and which could also be blamed for the widespread incidence of dental caries, especially amongst children. Attention soon turned to the amount of fat in the diet. In the United Kingdom, by the 1960s the medical profession was in general agreement that the rising fat content of the diet was contributing to the increase in coronary disease. Some debate ensued on the relative culpability of animal fats which were 'saturated' and plant and fish oils which were largely composed of 'unsaturated' fats. It led to an advertising war between margarine and butter producers and shattered the conventional view held for generations that animal foods – meat, eggs, and dairy produce – were 'protective'.[13] It also meant that the last quarter of the twentieth century was notable for the growth of a 'health foods' industry which generated concern over the inadequate control of labelling claims to promote health.

Contributors to this volume have identified an increasing tendency for populations to become overweight or obese in the last quarter of the twentieth century. In Britain, a longitudinal birth cohort study of children born in 1958 has provided measured obesity rates for almost 8,000 children who had reached the age of 33 years in the early 1990s.[14] By that age, 12.9 per cent were already obese. While research had previously suggested that poor cognitive function in childhood has been associated with obesity and type 2 diabetes in adults, this study claimed that poorer physical control and coordination in childhood also has an association with obesity in adults.[15] Such evidence reinforces the now generally held view that obesity is unequally distributed between and within countries: 'in affluent societies excess weight is more common among socially disadvantaged groups but the reverse is true in low income countries'.[16] Indeed medical management of obesity has recently looked beyond clinical treatment to 'the causes of obesity – genetic, hormonal, and environmental'.[17] Trying to solve the problems of obesity from a wider perspective than pharmaceutical or surgical approaches brings medical practitioners to behavioural, nutritional and physical exercise solutions. Epidemiologists have been amongst the first to note the extent to which the normalization of being overweight or obese is taking place, as there is now evidence that heavier people are 'less likely to think that they were overweight in

---

13   See Oddy 2003: 210–11.
14   *British Medical Journal* 2008: 337, a699. Measurements were based on BMI >30.
15   *British Medical Journal* 2008.
16   *British Medical Journal* 2007, 335: a347.
17   *The Lancet* 2003: 362, 1085.

2007 than in 1999'.[18] Such rationalization of body-size image indicates a growing resistance to any modification of eating habits, in particular by restricting intakes of alcohol, snacks, fast-foods and ready meals.

Policymakers have little to offer, because prescriptive measures will arouse opposition from vested interests such as the food industries. The European Commission has, in theory, promoted a campaign targeting children to reduce obesity by 'healthy' eating. Its Direction générale de la santé et des consommateurs has created the Sanco de la Commission Européenne, by sponsoring the French programme Ensemble Prévenons l'Obésité des Enfants (EPODE) which started in 2004 and operates in over 160 towns in France. EPODE was extended to Belgium as VIASANO in 2007 and to Spain as THAO Salud Infantil.[19]

Ultimately, the problem of excess bodyweight and obesity comes down to an energy equation: if energy intake is greater than energy expenditure, the excess will be laid down as adipose tissue. Whilst this volume has shown the effects of that imbalance in various European countries over the last hundred years, it will be up to future historians of diet to record and analyse how the resulting obesity epidemic progresses.

References

Branca, F., Nikogosian, H. and Lobstein, T. (eds), *The Challenge of Obesity in the WHO European Region and the Strategies for Response*, Geneva, 2007.

Comptroller and Auditor General, Report, *Tackling Obesity in England*, HC220, Session 2000–1, London, 2001.

Davidson, S., Pasmore, R., Brock, J.F., Truswell, A.S. *Human Nutrition and Dietetics*, Edinburgh, 1975.

European Commission, *Eurostat*, 2008.

Food and Agriculture Organization 'Committee on Calorie Requirements, 1950', *FAO Nutritional Studies* No. 5, Rome.

Food and Agriculture Organization 'Second FAO Committee on Calorie Requirements', 1957, *FAO Nutrition Studies* No. 15, Rome.

Food and Agriculture Organization Protein 'Requirements Report of the FAO Committee, 1957', *FAO Nutritional Studies* No.16, Rome.

---

18   *British Medical Journal* 2008: 337, a347, citing *British Medical Journal* 2008: 337, a494. There is as yet no evidence to support the perception that clothing manufacturers have increased the size dimensions of their women's garments.

19   *British Medical Journal* 2007: 335, 1238. See also www.epode.fr. The European network is a 'partenariat public / privé' which has attracted the food industry to its membership: Ferrero, Mars and Nestlé have joined. Nestlé France's slogan for its partnership is 'Manger bien pour vivre mieux'. In addition, the Belgian programme is supported by Carrefour and Unilever.

Food and Agriculture Organization 'Protein Requirements. Report of a Joint Food and Agriculture Organization FAO/WHO Expert Group', *FAO Nutrition Meetings Report Series* No. 37, Rome, 1965.

Merta, S. 'Karlsbad and Marienbad: the Spas and their Cures in Nineteenth-Century Europe', in ICREFH VIII, 2005, 152–63.

Oddy, D.J. *From Plain Fare to Fusion Food*, Woodbridge, 2003.

Office for National Statistics, *United Kingdom National Statistics*, London, Annual.

World Health Organization, *The SuRF Report 2. Surveillance of Chronic Disease: Risk Factors. Country-Level Data and Comparable Estimates*, Geneva, 2005.

World Health Organization, *Global Database on Body Mass Index*. Available at: http://www.who.int/bmi/index.jsp [accessed 20 April 2009].

Index